BIOMEDICAL PHYSICS AND BIOMATERIALS SCIENCE

The MIT Press
Cambridge, Massachusetts, and London, England

BIOMEDICAL PHYSICS AND BIOMATERIALS SCIENCE

Edited by
H. Eugene Stanley

Library of Congress Cataloging in Publication Data

Stanley, Harry Eugene.
 Biomedical physics and biomaterials science.

 Based on a summer program presented at M. I. T. in June 1971,
under the auspices of Harvard-M. I. T. Program in Health Sciences
and Technology.
 1. Biological physics. 2. Medical physics.
I. Massachusetts Institute of Technology. II. Title.
[DNLM: 1. Biomedical engineering. 2. Biophysics.
QT 34 S787b 1972]
R895.S69 574.1'91 72-8855
ISBN 0-262-19104-0
      0-262-69038-1 (pbk)

CONTENTS

Contents vi

PUBLISHER'S NOTE

The aim of this format is to close the time gap between the preparation of certain works and their publication in book form. A large number of significant though specialized manuscripts make the transition to formal publication either after a considerable delay or not at all. The time and expense of detailed text editing and composition in print may act to prevent publication or so to delay it that currency of content is affected.

The text of this book has been photographed directly from the author's typescript. It is edited to a satisfactory level of completeness and comprehensibility though not necessarily to the standard of consistency of minor editorial detail present in typeset books issued under our imprint.

The MIT Press

Numbers in parentheses indicate the pages on which the authors'
contributions begin.

ERNEST G. CRAVALHO, Ph. D. , Associate Professor of Mechanical
Engineering, M. I. T. (207).

ALVIN ESSIG, M. D. , Associate Professor of Medicine and Physiology, Tufts University School of Medicine; and Lecturer in Biophysics, Harvard Medical School (25).

JUDITH HERZFELD, Ph. D. , Research Associate, Harvard-M. I. T.
Program in Health Sciences and Technology (65).

CHARLES E. HUGGINS, M. D. , Associate Professor of Surgery,
Harvard Medical School; Chief, Surgical Low Temperature Unit,
Massachusetts General Hospital (191).

KEITH H. JOHNSON, Ph. D. , Associate Professor of Metallurgy
and Materials Science, M. I. T. (85).

JOHN G. KING, Ph. D. , Professor of Physics, M. I. T. (103).

THOMAS J. LARDNER, Ph. D. , Associate Professor of Mechanical
Engineering, M. I. T. (321).

PADMAKAR P. LELE, M. D. , D. Phil. Oxon., Professor of Experimental Medicine, Department of Mechanical Engineering and Department of Nutrition and Food Science, M. I. T. (257).

GEORGE W. PRATT, Ph. D. , Professor of Electrical Engineering,
M. I. T. (301).

ERIC L. RADIN, M. D. , Assistant Professor of Orthopedic Surgery,
Harvard Medical School; Lecturer in Mechanical Engineering,
M. I. T. (153).

ROBERT M. ROSE, Sc. D. , Professor of Metallurgy and Materials
Science, M. I. T. (169).

KENNETH J. ROTHSCHILD, Predoctoral Fellow, Department of
Physics, M. I. T. and Harvard-M. I. T. Program in Health Sciences
and Technology (3).

H. EUGENE STANLEY, Ph. D. , Associate Professor, Department of
Physics and Division of Health Sciences and Technology, M. I. T.(3, 65).

LLOYD V. SUTFIN, D. D. S. , Research Associate, Department of
Orthopedic Surgery, Harvard Medical School at the Children's
Hospital Medical Center (121).

JAMES C. WEAVER, Ph. D. , Research Associate, Physics Department, M. I. T. (103).

IOANNIS V. YANNAS, Ph. D. , Associate Professor of Mechanical
Engineering, M. I. T. (41).

PREFACE

A special summer program entitled <u>Biomedical Physics and Bio-materials Science</u> took place at the Massachusetts Institute of Technology in June 1971 under the auspices of the Harvard-M.I.T. Program in Health Sciences and Technology (Irving M. London, Director). Its purpose was to broaden contacts among workers in this increasingly important interdisciplinary field, and the unifying theme was the understanding and interpretation, on a physical level, of various biological problems of medical concern. Although the M.I.T. summer program represented only a survey of examples of interdisciplinary research, many participants felt that a collection of the lectures presented would be of interest to a wider group.

The editor wishes to thank Mrs. Carole Solomon for undertaking the task of preparing the camera copy and for carrying through her work with diligence and efficiency. Kenneth J. Rothschild kindly prepared the index.

H. Eugene Stanley
Director of the Program

PART I

FROM MOLECULE TO FUNCTIONING SYSTEM

CHAPTER 1

GLOBULAR MEMBRANE PROTEINS AS FUNCTIONAL UNITS OF IONIC TRANSPORT

Kenneth J. Rothschild and H. Eugene Stanley

INTRODUCTION

The selectivity of biological membranes remains one of the outstanding problems to be solved in biology. Clearly, if biological membranes were unable to selectively regulate which molecules pass through them, life as we know it would be impossible. In fact, one of the crucial steps in the evolution of life may have been the appearance of membranes which were able to select certain molecules from the external environment and concentrate them into the cell [1].

The cells found on earth today possess membranes which can exert highly selective control over the flow of molecules. For example, it is the cellular membrane which is responsible for the high concentration of K^+ found in most cells [2]; the molecular basis of this mechanism which enables the membrane to cause K^+ to be concentrated within the cell remains a mystery. Another example of the membrane's selective behavior is the ability of the nerve membrane to rapidly alter, with great precision, its permeability to Na^+ and K^+, thereby producing the well-studied nerve impulse [3].

In general, the transport function of the cellular membrane is two-fold. The membrane must block the entry of molecules which might be disruptive to the normal cell metabolism, and at the same time it must supply the basic molecules which are fuel for the cell's biochemical processes. Moreover, to maintain optimum biochemical conditions in the cell, it is also necessary that the levels of critical substances are regulated so that they remain fixed at predetermined concentrations. It appears that to meet this requirement, the cell membrane has developed, in many cases, a type of homeostatic control mechanism. For example, it is observed that increasing the concentration of Na^+ inside human erythrocytes leads to an increase in the "active transport" of these ions out of the cell, thereby tending to stabilize the internal Na^+ concentration at a fixed value [4:214].

In light of these facts, the cell membrane cannot be simply regarded as a "porous bag"; allowing some molecules to pass through it, and preventing larger molecules from passing. Significantly, the precision control of the molecular flow through the membrane is just as important to the survival of the cell as the control over the flow of molecules into and out of biochemical pathways in the cell. Both of these controls may be very similar on a molecular level, as we shall discuss in the next three sections.

GLOBULAR PROTEINS VERSUS LIPIDS AS FUNCTIONAL UNITS OF TRANSPORT

Singer [5] has recently proposed a structural model of the biological membrane which offers an extremely simple way to account on a molecular level for a wide variety of membrane functions. Singer pictures a membrane consisting of globular proteins embedded in a relatively fluid lipid bilayer. At least some of these proteins are postulated to extend completely through the bilayer, while others are on the surface or partially embedded in the membrane (c f. Fig. 1). A number of experiments lend convincing support to this picture [5, 6].

The observation that globular proteins are dissolved into the mem-

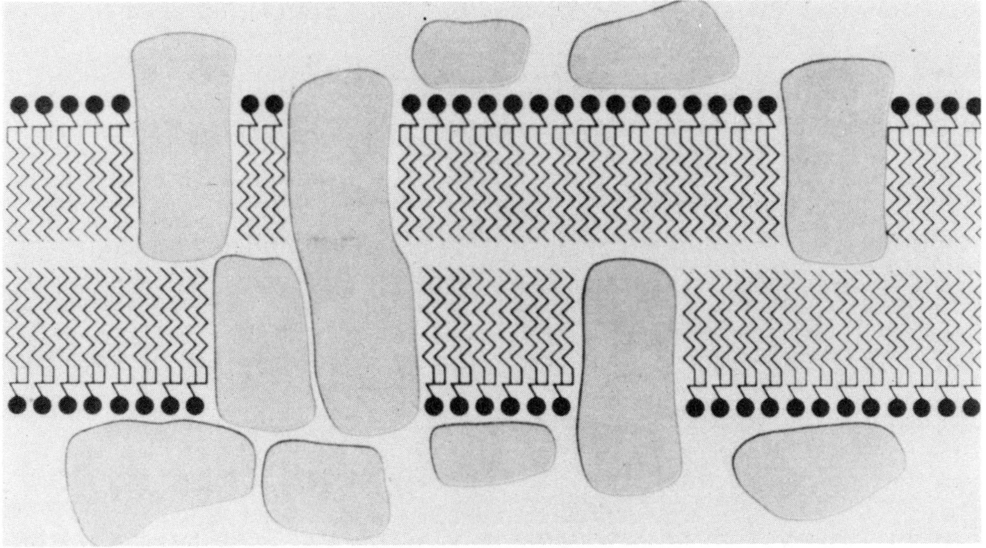

Figure 1.   Singer model of membrane structure. Proteins exist pre-dominantly in globular configuration and may extend across bilayer[6].

brane rather than extended in a beta configuration on the surface [7:1] or intertwined with the lipids [8:265] is very significant. Specifically, since globular proteins have been found to be the functional units in so many other places inside the cell [9], it is tempting to conjecture that some globular proteins also serve as functional units of transport in biological membranes. In this context, the lipid bilayer, originally proposed to be the basic structure of the membrane by Davson and Danielli [10], can be thought of as a pudding in which are inserted the functional plums (i.e., globular proteins). Many transport properties of the membrane will then be largely determined by the kind of membrane proteins coded for by the cell's genome.

It should be pointed out, however, that this hypothesis does not imply that membrane proteins are responsible for the transport of all substances. It is well known that the lipid composition of the cellular membrane has a great influence on the membrane's selectivity to some permeants. For example, in 1900 Overton observed [11:88] that a definite relationship exists between the permeability of lipid-soluble substances through biological membranes and their water-lipid partition coefficients (cf. Fig. 2). For this reason, it is expected that non-electrolytes permeate through the lipid bilayer part of the membrane without the aid of the membrane proteins.

However, if we postulate that the lipids can be arranged in such a way as to account for the transport of electrolytes (e.g., ions), we are confronted with the following difficulties:

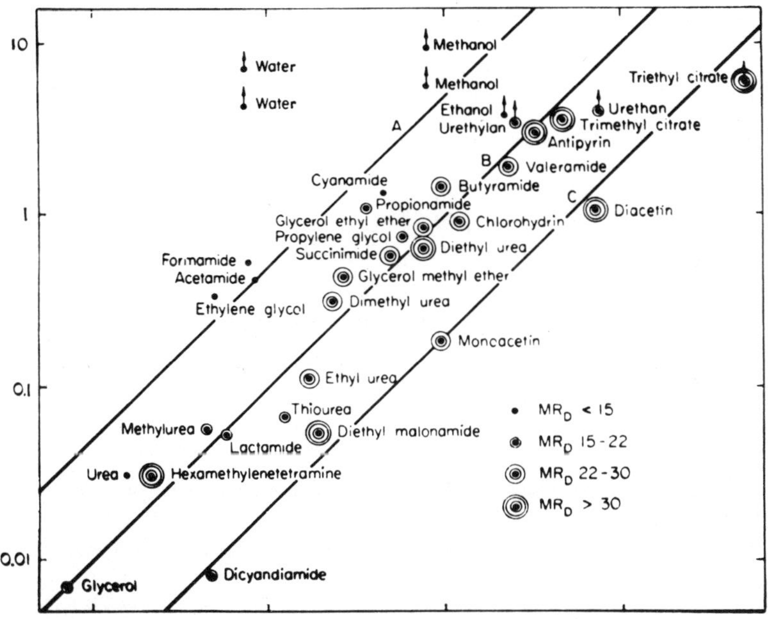

Figure 2.   Relation between permeability of non-electrolytes through Clara membrane and oil-water partition ratio. Abcess; permeability x $M^{1/2}$, where M is molecular weight of substance. Ordinate, liquid-water partition ratio [34].

(1) Biological membranes often have a conductivity as much as a million times greater than synthetic lipid bilayers [12:59]. Cyclic depsipeptides, such as valinomycin, when inserted into synthetic lipid bilayers, produce a dramatic increase in the selective permeability of these membranes to certain ions. Further it is thought some of these molecules achieve this increase in membrane permeability by forming stacked pore arrangements [13]. Without these agents, the lipid bilayer is highly impermeable to electrolytes.

(2) There is no obvious way in which phospholipids can be arranged so that they can control the transport of electrolytes and particularly ions. It has been suggested [14:233] that micellar formation may lead to the formation of charge-lined pores. However, selective control over permeability implies the existence of specific types of micelles composed of special lipids which have segregated from the total population of phospholipids. The self-assembly of different specific pore structures out of a sea of phospholipids is unlikely, since phospholipids are not molecules that contain a high degree of structural information.

(3) Phospholipids are not designed for the exquisite regulatory functions of which other macromolecules (e. g. , proteins and nucleic acids) seem capable. Thus, rapid changes of membrane permeability during nerve excitation, and coupled active transport of ions, seem more suited for control by protein than for control by phospholipid. Moreover, induction of selective transport of a normally non-permeating molecule is observed to occur in some cells under certain conditions. This is more likely to be accomplished by the synthesis of a new membrane protein rather than by an alteration of the lipid composition of the membrane [15:632].

CONTROL OF IONIC TRANSPORT BY GLOBULAR PROTEINS

In this section we shall discuss how globular proteins in the mem-

brane may control ionic transport. Although a detailed picture of
how membrane proteins control ionic transport will not be available
for a long time, it is not unreasonable to suspect that certain general
mechanisms exist in membrane proteins, just as some general mechanisms have been found in enzymes (e. g. , allostery). Thus, one of
the motivations for what follows is the hope that the examination of
specific examples of ionic transport in biological membranes may
shed light on the functional properties of some globular membrane
proteins.

Two general models that have been proposed previously to account
for the transport of ions by proteins are the diffusing carrier model
[16:163] and the rotating carrier model [17:87]. Such proteins have
been referred to elsewhere as permeases [34]. We will not discuss
these models here, but we note that one major objection voiced
against these models is the radical departure they represent from the
way proteins are known to function elsewhere in the cell.

A third picture will now be discussed [18]. It is called the _permion_
model, where permion refers to those globular proteins which extend
across the membrane and which are involved in membrane transport.
The two major features of the permion model (cf. Fig. 3) are the
following:

(i) The membrane protein contains a channel,

(ii) The membrane protein may undergo a quarternary conformational
transition between two states resulting in the movement, relative to

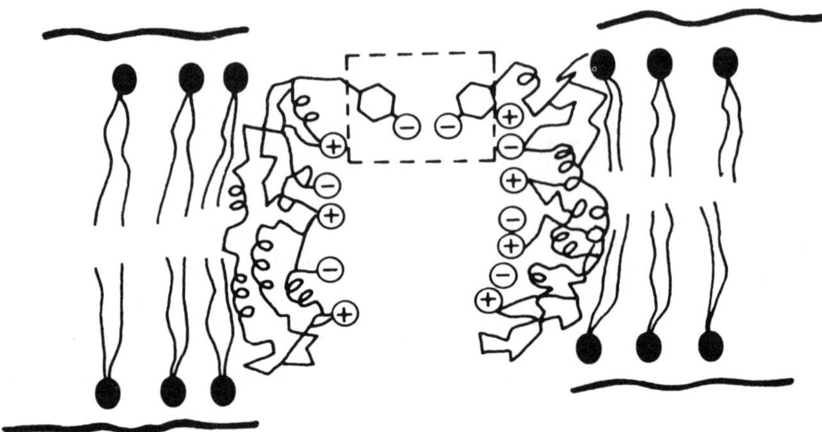

Side view

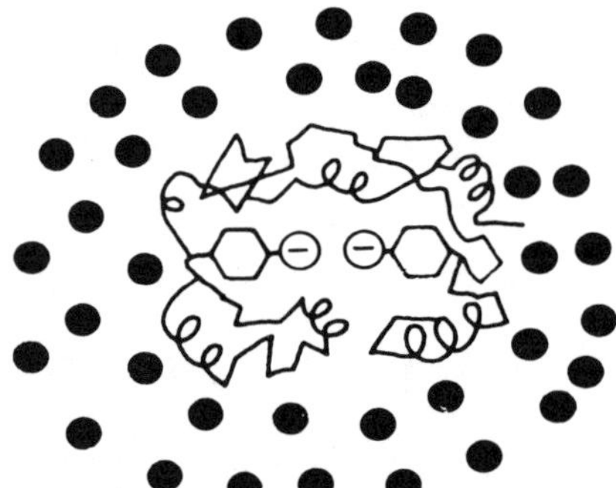

Top view

Figure 3.   (a) Side view and (b) top view of a multi-subunit globular protein dissolved in a lipid sea.   The protein ('permion') contains a water-filled channel and a barrier ('control site') consisting of two negative residues.

each other, of a small number of the polar residues which are direct-

ly involved in the control of transport.

Although the postulates (i) and (ii) will remain conjectural until
more experimental work is done on membrane proteins, both of these
features have been found separately in other proteins.

(i) Proteins with channels: The multisubunit proteins pyruvate carboxylase (Fig. 4) and glutamine synthetase have been shown by high-
resolution electron microscopy to contain channels [19:165]. In addition, Perutz [20:113] has demonstrated by x-ray analysis a 10Å
channel in the four-subunit protein hemoglobin. This feature may

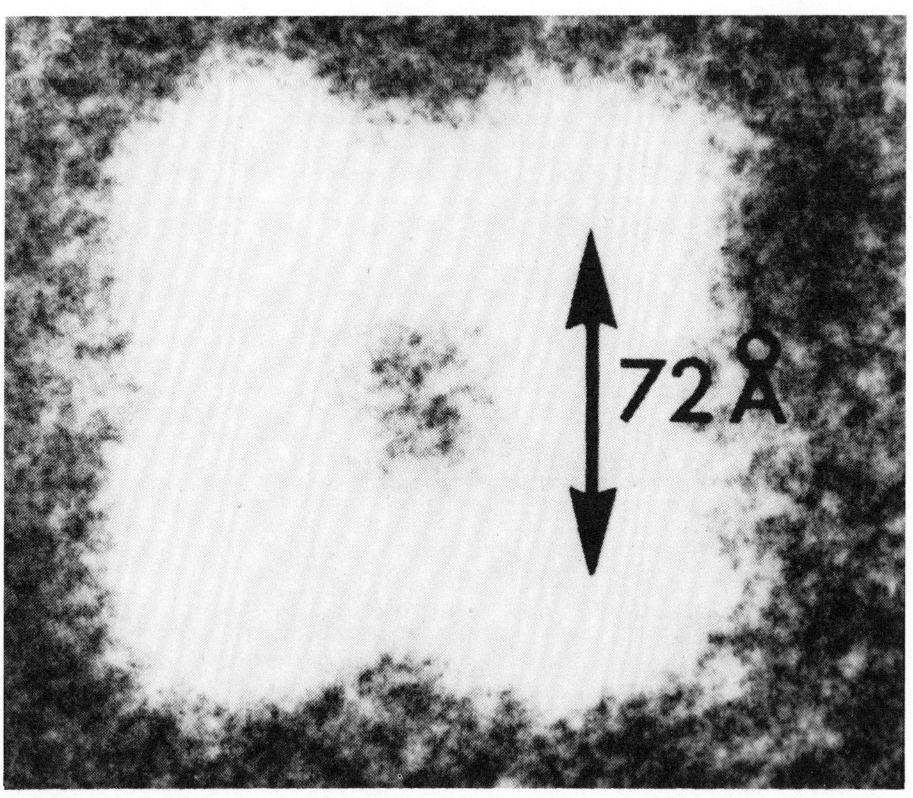

Figure 4.   High resolution electron micrograph of the enzyme pyruvate carboxylase, which appears to contain four subunits and a large
central core [19].

well be found to be quite common, as the detailed 3-dimensional structures of more multi-subunit proteins become available. It has also been argued [18] that it may be energetically favorable for channels to form inside membrane proteins.

(ii) Quarternary conformational transitions in proteins: Quarternary transitions between two conformations of an enzyme are believed [21:88] to be an important mechanism for the overall control of biochemical reactions. These allosteric proteins work in the following way. A binding ligand acts to modify the enzyme structure, thereby affecting catalysis. Further, since the binding ligand is often a product of some part of the biochemical pathway that is under control of this particular enzyme, the enzyme is in fact under feedback control. This allosteric mechanism of achieving precision in enzyme control may be analogous to the mechanism of achieving precision in membrane function [22:335].

APPLICATIONS OF THE PERMION CONCEPT

A specific suggestion for a molecular model which utilizes only the above two hypothesized properties of a membrane protein, and which is capable of accounting for a wide range of different types of ionic transport has been proposed [18]. The model is briefly described below, along with its application to some of the different types of ionic transport observed in biological membranes.

In Fig. 5 we illustrate the major features of this model. They are

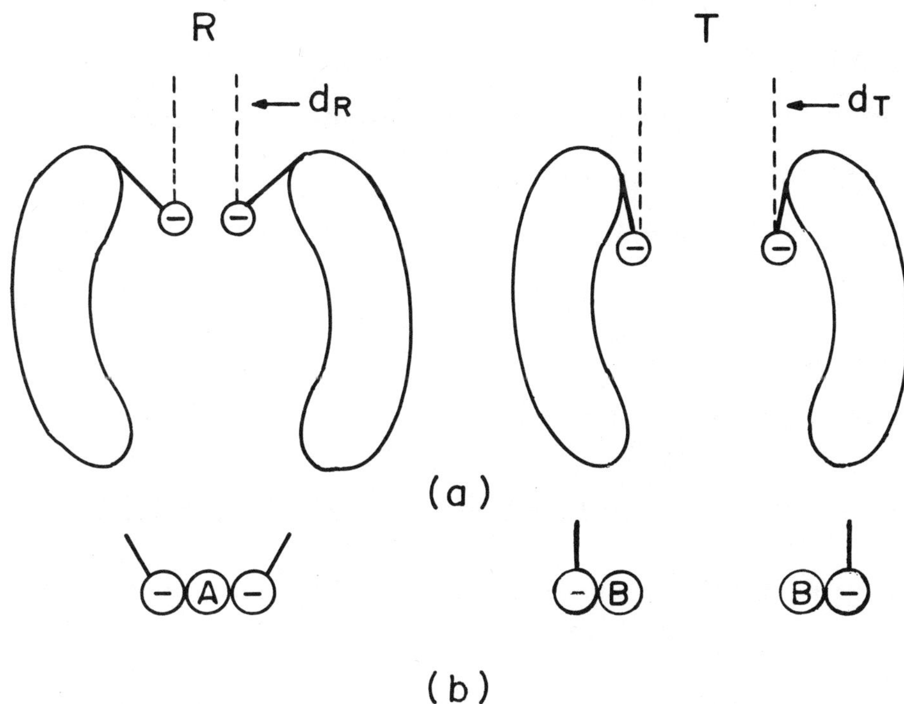

Figure 5. (a) Two possible permion conformations, R and T, along with (b) Binding of two different cations, A and B, to the R and T conformations.

the following:

(1) A transition between two quarternary conformations, R and T, of the membrane protein alters the distance, d, between two polar residues.

(2) This change in d alters the selectivity which the two residues have for different ions. For example, we show in Fig. 5 that a trans-

ition from R to T can lead to an exchange of one A cation for two B cations. According to this principle, A will tend to stabilize the protein in conformation R, while B will stabilize the protein in conformation T. E. g., for nerve membrane, A might be Ca^{++} and B might be K^+. A theoretical discussion (based on the work of Eiseman [23:259] and Ling [24])of ionic association with fixed residues (along with a derivation of this predicted effect) is given elsewhere [25].

(3) Some ions cannot pass through the permion in conformation R, but can in conformation T.

Utilizing this model we are able to qualitatively account for several different types of ionic transport.

Simple Passive Transport with Linear Dependence on Concentration (cf. Fig. 6a)

This type of transport will result if the probability at a given time of a cation passing from side I of the membrane to side II is proportional to the concentration of the cation on side I (and independent of the concentration on side II).

In terms of the model presented, if the permion's conformation is independent of the permeating species and there exist enough open channels so that saturation does not become important at high concentrations of cation, the dependence of J on C will be linear (cf. Fig. 6a). When saturation becomes important, we expect a leveling off of the plot as indicated by the dotted line. One example of this type of kinetics is the efflux of potassium in lead-poisoned red cells ghosts [26:319].

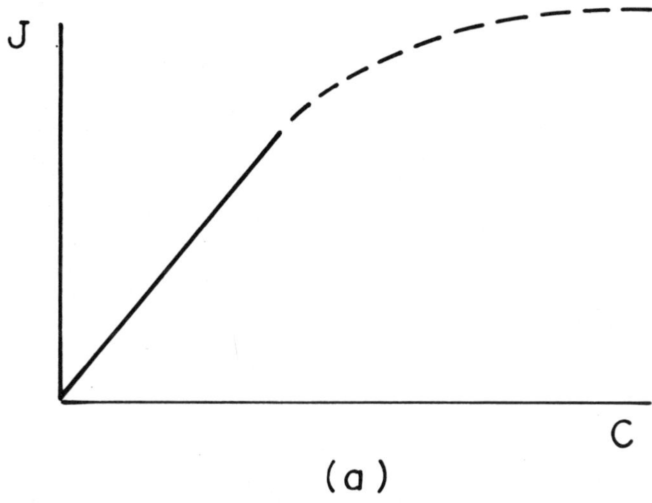

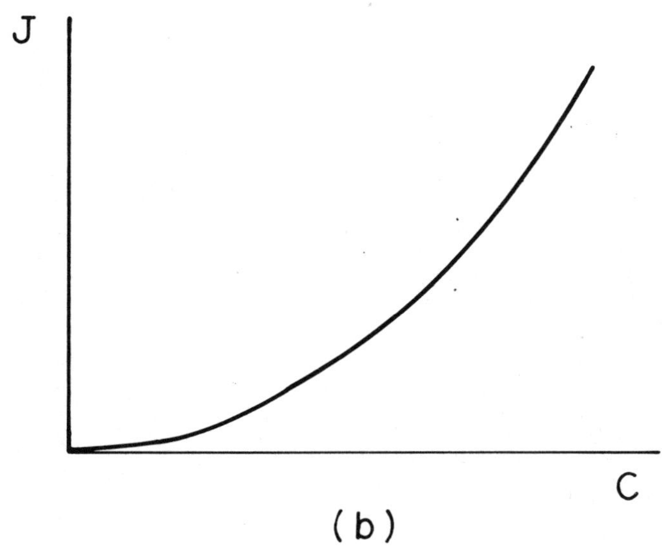

Figure 6. (a) Solid line - simple passive transport with linear dependence of flow, J, on concentration, C. Dotted line - effect of saturation in high concentration region. (b) Passive transport with higher order dependence on concentration.

Passive Transport with Higher Order Dependence on Concentration
(cf. Fig. 6b)

This type of transport will result if the probability of a permion open-

ing into a conformation which can pass an ion depends on the binding

to the permion of one or more ions of the same species. For example,

in Fig. 7a we show on a molecular level how the binding of cation B

can stabilize the permion in an open conformation. Thus, the pro-

bability of a B cation passing through a permion at a given time de-

pends on the probability of two other B cations binding. The depend-

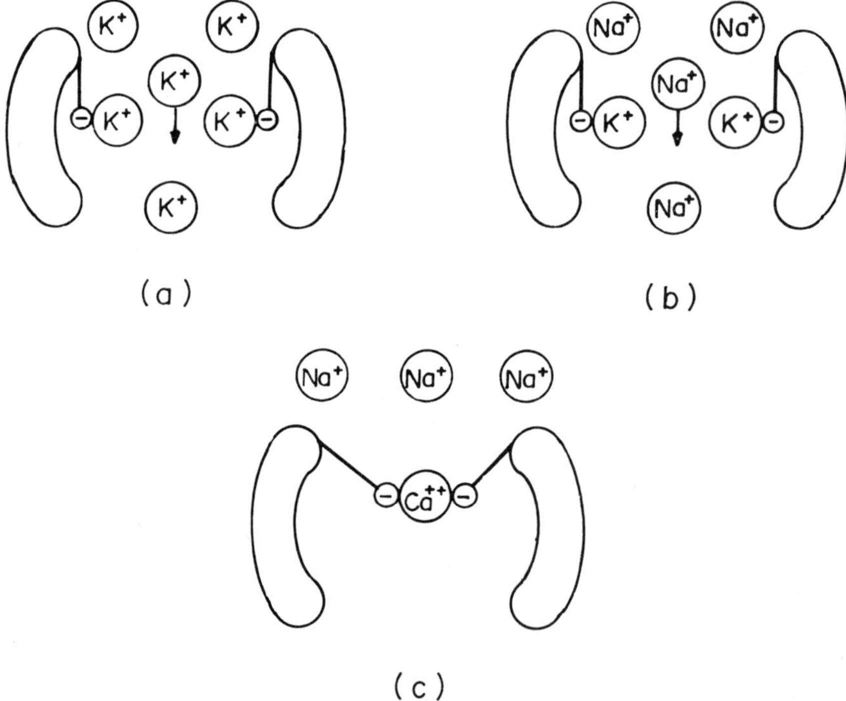

Figure 7.   (a) Permion passing B cation (K^+), stabilized by two
other B cations. (b) Permion passing C cation (Na^+), due to stabilization in the T conformation by two B cations (K^+). (c) Permion
being stabilized in closed conformation by the A cation (Ca^{++}) preventing passage of C cation (Na^+).

ence of J on C will then be higher than first order. The influx of K^+ into the giant nerve fiber from sepia and the efflux of K^+ from the squid axon seem to obey this type of kinetics [27:61, 28:534].

Passive Transport with Dependence on Second Ionic Species

It is often observed that the probability of a cation passing from side I of the membrane to side II depends on the existence of a second ionic species [26:319, 29:24]. In terms of the permion model, if a specific cation, B, is required to stabilize the permion in an open conformation (cf. Fig. 7b), the flow of the species C may depend not only on its own concentration, but also on the concentration of species B. The possibility exists also that another ion might lower the probability of C passing by tending to stabilize the permion in a closed conformation (cf. Fig. 7c).

It has been postulated [18, 32] that this situation may be present for the Na^+-channels in nerve membrane. Here, K^+ is required to open the Na^+-channel, displacing Ca^{++} and therby allowing Na^+ to pass (cf. Fig. 7b).

Coupled Active Transport of Na^+ and K^+

The coupled active transport of Na^+ and K^+ has been observed in a variety of biological membranes [30:596]. If we follow the suggestion of Jardetsky [31:969] and modify the permion picture (cf. Fig. 5) so that in the R conformation the channel is closed to the intracellular medium and that in the T conformation the channel is closed to the extracellular medium, we can then account for the coupled transport

of Na^+ and K^+ by assuming that

1. A cation = K^+,
2. B cation = Na^+.

With these assumptions we observe that as the permion switches

between the R and T conformation (due somehow to an input of met-

abolic energy)   two Na^+ ions will be transported into the extracellu-

lar medium for each K^+ transported into the intracellular medium

(cf. Fig. 8).

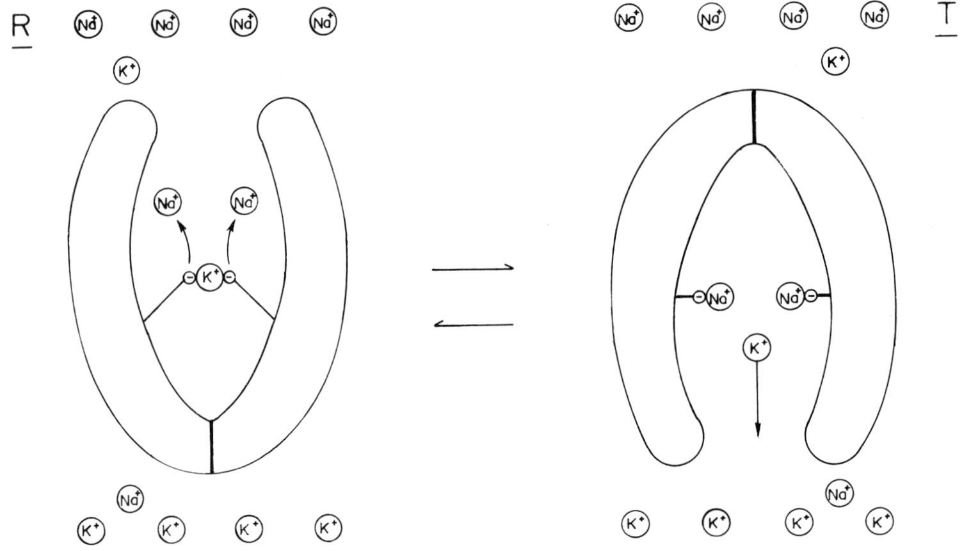

Figure 8.   Coupled active transport of Na^+ and K^+ by permion switch-
ing between R and T conformations.

It is important to note that in order for this model to work, the permion channel must be small enough to exclude all but the bound ions during its transition from the R to T conformation. A similar model, which can account for a $3Na^+/2K^+$ active transport has also been proposed [33].

CONCLUSION

There is evidence [5:146] that some membrane proteins exist in a globular conformation and extend completely across the lipid bilayer. In view of this, the hypothesis that some membrane proteins may have evolved the functional capacity to control ionic transport in biological membranes is worth consideration.

We have discussed in this paper how a globular protein might function to control ionic transport. Conformational changes and the related binding of ions to a few key polar residues are postulated to be two important features of a membrane protein which controls ionic transport. With these two properties, it is shown how different types of ionic transport, including Na^+-K^+ active transport and changes in the permeability to Na^+ in the nerve membrane, might result.

The extension of these ideas to the selective transport of electrolytes such as glucose can be easily made. A membrane protein would have a binding site for glucose which is accessible from either side of the membrane depending on the conformation of the protein. Transport of glucose across the membrane would then depend on the "glucose transport protein" undergoing a conformational change.

Both the active and passive transport of glucose could be achieved

using this mechanism, which is quite different from the carrier model

discussed elsewhere [16, 17:87].

<u>REFERENCES</u>

[1]   Calvin, M. 1969. <u>Chemical Evolution.</u> New York: Oxford University Press.

[2]   Lehninger, A. 1970. <u>Biochemistry.</u> New York: Worth Publishers.

[3]   Katz, B. 1966. <u>Nerve, Muscle and Synapse.</u> New York: McGraw-Hill Book Company.

[4]   Whitman, R. and M. E. Ager. 1965. The Connection between Active Cation Transport and Metabolism in Erythrocytes. <u>Biochem. J.</u> 97:214.

[5]   Singer, S. J. 1971. In <u>Structure and Function of Biological Membranes,</u> ed. L. I. Rothfield, p. 146. New York: Academic Press.

[6]   Fox, C. F. 1972. <u>Scientific American.</u> 226:30.

[7]   Robertson, J. D. 1964. In <u>Cellular Membranes in Development,</u> ed. M. Locke, p. 1. New York: Academic Press.

[8]   Benson, A. A. 1966. <u>J. Amer. Oil Chem. Soc.</u> 43:265.

[9]   Dickerson, R. E. and I. Geis. 1969. <u>The Structure and Action of Proteins.</u> New York: Harper and Row.

[10]   Davson, H. , and J. F. Danielli. 1962. <u>Permeability of Natural Membranes.</u> 2nd ed. London and New York: <u>Cambridge University Press.</u>

[11]   Overton, E. 1899. <u>Vierteljahressch. Naturforsch. Ges. Zurich.</u> 44:88.

[12]   Hanai. T., D.A. Hayden, and J. Taylor. 1965. _J. Gen. Physiol._ 48:59.

[13]   Eigen, M. and L. Demayer. 1969. _Neurosciences Research Program Bulletin._ Vol. 9.

[14]   Lucy, J.A. 1968. _Biological Membranes_, ed. D. Chapman, p. 233. New York: Academic Press.

[15]   Pardee, A.B. 1968. _Science._ 162:632.

[16]   Widdas, W.F. 1954. _J. Physiol._ 125:163.

[17]   Davies, R.E. and R.D. Keynes. 1961. In _Membrane Transport and Metabolism_, eds. A. Kleinzeller, and A. Kotyk, p. 87. New York: Academic Press.

[18]   Rothschild, K.J. and H.E. Stanley. 1972. In _Membranes, Viruses, and Immune Mechanisms in Clinical and Experimental Diseases_, eds. S. Day and R. Good. New York: Academic Press. To be published.

[19]   Valentine, R.C. 1969. In _Symmetry and Function of Biological Systems at the Macromolecular Level_, eds. A. Engstrom and B. Strandberg, p. 165. New York: Wiley.

[20]   Perutz, M.F. 1969. _Proc. Roy. Soc., Ser. B_, 173:113.

[21]   Monod, J., J. Wyman, and J.-P. Changeux. 1965. _J. Mol. Biol._ 12:88.

[22]   Changeux, J.-P., J. Thiery, Y. Tung, and C. Kittel. 1967. _Proc. Nat. Acad. Sci._ 57:335.

[23]   Eisenman, G. 1962. _Biophys. J._ 2:259.

[24]   Ling, G. 1962. _Physical Theory of the Living State._ New York: Blaisdell Publishing.

[25]   Rothschild, K.J. 1972. Proc. of the IV International Conf. on Biophysics, Moscow. To be published.

[26]   Passow, H. 1969. In _The Molecular Basis of Membrane Function_, ed. D. C. Tosteson, p. 319. New Jersey: Prentice-Hall.

[27]   Hodgkin, A.L. and R.D. Keynes. 1955. _J. Physiol._ 28:61.

[28] Sjodin, R.A. and L.J. Mullins. 1967. *J. Gen. Physiol.* 50: 534.

[29] Kasai, M. and J.-P. Changeux. 1971. *J. Memb. Biol.* 6:24.

[30] Skou, J.C. 1965. *Physiol.* Rev. 45:596.

[31] Jardetzky, O. 1966. *Nature.* 211:969.

[32] Rothschild, K.J. 1972. Ph.D. Thesis, M.I.T.

[33] Rothschild, K.J. and H.E. Stanley. 1972. *Nature* Submitted for publication.

[34] Cereijido, M. and C. A. Rotunno. 1970. *Introduction to the Study of Biological Membranes.* New York: Gordon and Breach Science Publishers.

GENERAL BIBLIOGRAPHY

In addition to Refs. [2], [5], [9], [10], [24], [26] above, please see

the following:

Kavanau, J. L. 1965. *Structure and Function in Biological Membranes.* Vols. 1 and 2. San Francisco: Holden-Day, Inc.

Snell, F., J. Wolken, G. Iverson, and J. Lam, eds. 1968. *Physical Principles of Biological Membranes.* New York: Gordon and Breach Science Publishers.

Bolis, L., A. Katchalsky, R. D. Keynes, W. R. Leowenstein, and B. A. Pethica, eds. 1969. *Permeability and Function of Biological Membranes.* Amsterdam: North-Holland Publishing Company.

Chapman, D. 1969. *Introduction to Lipids.* London: McGraw-Hill Publishing Company.

New York Heart Association. 1969. *Membrane Proteins.* Proceedings of Symposium. Boston: Little, Brown and Company.

Bittar, E. E., ed. 1970. *Membranes and Ion Transport.* Vols. 1, 2 and 3. London: Wiley Interscience.

Adelman, W. J., Jr., ed. 1971. Biophysics and Physiology of Excitable Membranes. New York: Van Nostrand Reinhold Company.

Porcellati, G. and F. diJeso, eds. 1971. Membrane-Bound Enzymes. New York: Plenum Press.

Szwarc, M., ed. 1972. Ions and Ion Pairs in Organic Reactions. New York: Wiley Interscience.

CHAPTER 2

SALT AND WATER TRANSPORT IN BIOLOGICAL SYSTEMS

Alvin Essig

A biological consideration of fundamental interest is that of homeo-
stasis, the ability of animals to maintain relatively stable internal
environments despite the vicissitudes of both health and disease.
Obviously, this is a very broad topic which can be fruitfully studied at
many levels and by many approaches. I would like to direct your
attention to the fluid which bathes the cells of the body, the extracellular fluid, and the mechanisms which account for the maintenance of
normal volume and electrolyte composition of this compartment [1].

As you may or may not realize, an impressively large fraction of
Homo sapiens is water; in the normal male water may constitute some
50-75 percent of total body weight; in females, who tend to be fatter,
the average figure is somewhat lower. Of the total water content,
about one third is outside cells, providing the route for delivery of
nutrients and other substances necessary for normal cellular function,
and for the removal of the end products of cellular metabolism.

The major components of the extracellular fluid are electrolytes,
predominantly salt, NaCl. Since Na^+ is present in high concentration
in the extracellular fluid and to a large extent excluded from the
interior of cells its osmotic activity makes it an important determinant
of extracellular fluid volume. Thus, when for some reason the body is

deficient in salt there is depletion of the extracellular fluid volume,
with harmful effects on the effectiveness of the circulatory system.
On the other hand, diseases such as congestive heart failure, or certain disorders of the liver or kidneys, often lead to the accumulation
of excess salt in the body, with resultant overexpansion of the extracellular fluid volume. Again this is a cause of discomfort and danger
to the patient.

As you may well imagine, since either contraction or overexpansion
of the extracellular fluid compartment may represent a grave threat,
the body is equipped with effective devices which tend to prevent these
developments. One important regulatory factor is the hormone
aldosterone. Normally, when an individual is made slightly deficient
in Na^+, whether as a consequence of diminished intake or excessive
losses, there is an increased release of aldosterone by the adrenal
gland. Aldosterone acts on the kidney, causing it to retain sodium
more avidly than usual. With a continuing dietary intake of salt there
is then a gradual increase in body Na^+ content, with eventual return
to normal. Contrariwise, excess Na^+ intake depresses aldosterone
release, leading to an increased rate of excretion of Na^+ by the kidneys.

Not only the quantity but also the composition of the body fluids is
important. In particular, here I refer to the total solute concentration
or _osmolality_ of the body fluids. Here again, deviations in either
direction, i. e. , either severe dehydration and concentration of the

body fluids, or excessive hydration and dilution of body fluids, can result in significant impairment of the function of several organ systems. Accordingly, again as you would expect, there are factors which act to prevent this state of affairs. The most important of these is the antidiuretic hormone (ADH). Normally, even slight water deficiency, with only a slight elevation in the concentration of the body fluids, stimulates the release of ADH by the posterior pituitary gland. As does aldosterone, ADH acts on the kidney. In this case the effect is to increase the reabsorption of water; with continuing water intake there is then gradual dilution of the body fluids toward normal. If, on the other hand, the individual has overdiluted his body fluids the release of ADH is repressed, facilitating the loss of excess water by way of the kidneys. I would point out that this, like the aldosterone mechanism, is an example of an extremely sensitive feedback mechanism; normally the ADH mechanism maintains the osmolality of the body fluids in a range of 285-295 milliosmols per liter.

In order to understand both the aldosterone and ADH mechanisms better it is worthwhile to spend a few minutes considering the structure of the kidney and the gross details of its handling of salt and water. Each human kidney, which may be said to be shaped like a kidney bean, contains some one million functional units or nephrons, as pictured schematically below. The kidneys are vigorously perfused; each minute about 1200 ml of blood, about a fourth of the total cardiac output, reaches the kidneys. Roughly a fifth of the fluid which is deliver-

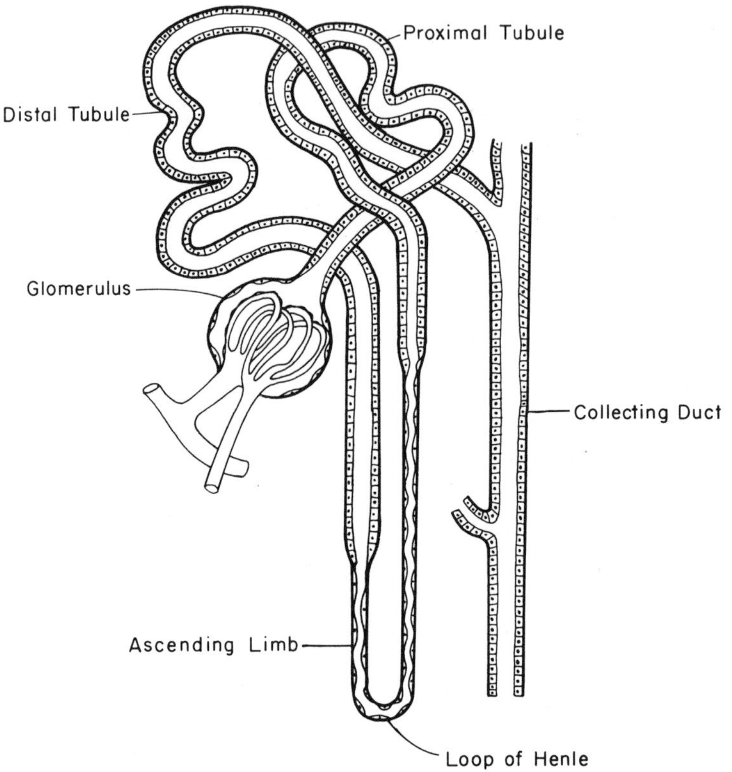

Figure 1.

ed to the kidney is filtered through the glomeruli into the interior of

the kidney tubules.   This might seem to be a means of effective remov-

al of waste products from the body, but there are important constraints

on the system.   For one, it is easily appreciated that, if the unaltered

filtrate were removed from the body at a rate approximating that of

glomerular filtration, within a very few minutes there would be a catastrophic depletion of body sodium and water content. Of course this does not occur. With respect to salt and water, kidney function is very roundabout, in that there may be a reabsorption of well over 99 percent of the quantities filtered.

The nature of sodium reabsorption in the kidney is of considerable interest, serving as an example of a process common to a great variety of plant and animal cells. Sodium transport is vigorous, occurring as a result of the expenditure of metabolic energy. This <u>active transport</u> takes place even in the face of adverse concentration and electrical potential gradients. That is, it can be shown that as a consequence of the expenditure of metabolic energy sodium can move in a direction opposite to that which would be seen in non-living <u>passive</u> systems. Furthermore, it has been demonstrated that in certain regions of the kidney the rate of active sodium transport is enhanced in the presence of aldosterone, providing an explanation for the ability of this hormone to regulate the sodium content of the body.

Now we would like to look at the other part of the story - the ability of the kidney to elaborate dilute urine when it is necessary to excrete extra water, and to form concentrated urine when it is necessary to conserve water. This ability depends on mechanisms involving both salt and water transport. Throughout much of the kidney, the region called the <u>proximal tubule,</u> the renal handling of sodium and water are closely linked, possibly as a consequence of the high premeability of the tissue to water. Without going into the mechanisms at this point we can say that vigorous sodium transport is necessarily associated with a brisk rate of water reabsorption, _pari passu_. This process accounts for the reabsorption of some 2/3-4/5 of filtered sodium and water. However, it cannot account for either the conservation of water stimulated by dehydration or the rapid elimination of water in response to overhydration, since in the proximal tubule solutes and water are reabsorbed in the same proportions as in the extracellular fluid. To explain the ability of the kidney to form either a dilute or concentrated urine we must consider a more subtle relation between salt and water transport. This depends not only on the sodium transport mechanisms we have already considered, but also on the pattern of water permeability in the kidney. I refer here to the fact that the permeability of the tubule walls to water varies in different regions of the kidney and varies with the conditions of the system. More specifically, there are regions of the kidney where water permeability is much lower than in the proximal tubule, and regions where the permeability varies accordingly as there is a need either to excrete or to conserve water.

The implications of a region of low water permeability are easily appreciated. In the region of the nephron just past the first bend, which we call the _ascending limb_ of the _loop of Henle_, as elsewhere in the kidney, a vigorous active transport mechanism moves sodium out of the tubular fluid. Here, however, in contradistinction to the proximal tubule, water permeability is low. As a consequence, removal of sodium results in the formation of a dilute tubular fluid and a concentrated interstitial compartment.

What happens to the tubular fluid past this point reflects the other aspect of the water transport mechanism which I have mentioned, the variability of water permeability under different conditions. More specifically, when there is a need for the kidneys to eliminate excess water the water permeability of the distal tubule and collecting ducts is low. Hence, despite the presence of concentrated interstitial fluids, which would tend to abstract water from the interior of the tubule, the tubular fluid remains dilute as it moves through the kidney, with the resultant elaboration of dilute urine. On the other hand, when there is a need for the body to conserve water, the situation is very different. Under these circumstances the pituitary gland releases anti-diuretic hormone (ADH), which increases the water permeability of the distal tubules and collecting ducts. Accordingly, there is a tendency for water to move across the tubule wall out of the dilute tubular fluid into the concentrated interstitial fluid. Since the interstitial fluid is increasingly concentrated descending from the outer cortical region to the inner medullary region, the fluid becomes increasingly concentrated as it moves down through the collecting duct, resulting in the formation of a highly concentrated urine.

The information which I have just given you concerning the overall character of the kidney's regulation of salt and water excretion represents understanding based on literally decades of work employing a host of experimental techniques. Some of these have dealt with intact animals, and others with perfused isolated organs, or even isolated kidney tubules. In recent years a great deal of work has employed the technique called micropuncture. In this approach a kidney of an anesthesized animal is exposed, partially freed from its attachments, and secured. With the aid of a dissecting microscope the experimenter then places micropipettes, and in some cases a microelectrode, at accessible sites within the lumen of a kidney tubule. It is then possible to sample tubular contents and to evaluate the flow, the hydrostatic pressure, and electrical parameters, while altering some experimental variable, for example by the administration of electrolyte solutions, a drug, or a hormone. As you heard, these various methods have given renal physiologists and clinicians a good deal of insight into the workings of the kidney, but you can appreciate that one cannot expect to unravel fine details of mechanism by means of these gross techniques. For these purposes it is important to control the environment of experimental tissues more completely, and thereby improve the precision of observations of local events.

Fortunately, there are two relatively simple tissues which serve as useful model systems for the kidney, namely the frog skin and the toad bladder [2:110] [3:657]. Each of these tissues shows important parallels with kidney function with respect to active sodium transport, permeability to water, and response to the administration of the hormones aldosterone and ADH. Each tissue is a flat thin membrane, and is readily studied while mounted between two halves of a lucite chamber. If both surfaces are bathed by an appropriate electrolyte solution the tissue remains viable for several hours. With the aid of appropriate bridge-electrode systems one can monitor the electrical potential difference across the membrane and pass sufficient current to set the potential at the level desired. It is also possible to measure material flows across the membrane. Solute flows are readily determined by tracer isotope techniques.

When a frog skin or toad bladder is mounted in this arrangement and exposed to identical physiological electrolyte solutions at each surface it is found that the inner (or body-side) surface is electrically positive relative to the outer surface. This is the consequence of active sodium transport. Somehow energy derived from the metabolism of tissue substrates brings about the net transport of sodium ions from the outer to the inner bathing solution; the associated displacement of net charge may result in electrical potential differences of some 10-100 mV. When external circuitry is used to fix the electrical potential difference at zero one can measure a current generated by the tissue. For appropriate species this <u>short-circuit current</u> is electrically equivalent to the rate of net sodium transport, indicating that sodium is the only ion which is being actively transported. This is a great convenience, since it permits the use of an ammeter to monitor the rate of active sodium transport, and to observe promptly the effects of drugs, hormones, or changes in the bathing solutions. For example, within two to three hours after exposure to 10^{-7} M aldosterone the short circuit current rises markedly, with a corresponding change in the rate of active sodium transport [4:1185].

Because of the experimental convenience of this technique active sodium transport in the frog skin and toad bladder has been studied intensively from several points of view -- anatomical, physiological, pharmacological, biochemical, and biophysical. There is not time to present even a summary of what has been learned by these means. Instead I shall tell you a bit about some studies involving the application of non-equilibrium thermodynamics. The background to this is the extensive work of physicists and physical chemists on the nature of energy conversion in non-living systems. My colleagues and I have felt that the application of non-equilibrium thermodynamics provides insights into the mechanisms of active sodium transport which would not be readily obtained by the more standard biological approaches [5:1434].

Consider a frog skin or toad bladder exposed to identical solutions at each surface, so that there is no concentration gradient influencing transport. We would expect the rate of sodium transport J_{Na} to be influenced by the electrical potential difference $\Delta \psi$ across the membrane, but since sodium transport is "driven" by metabolism we would expect it also to be influenced by the free energy of the metabolic driving reaction ΔF. Similarly, we expect the rate of metabolism J_r to be determined largely by ΔF, but since transport and metabolism are coupled, we expect J_r to be influenced also by $\Delta \psi$. There is a range in which these dependencies will be linear; we anticipate that this range may be sufficiently large to be experimentally useful. Accordingly we write

$$J_{Na} = L_{Na}(-\mathscr{F}\Delta\psi) + L_{Nar}(-\Delta F) , \tag{1}$$

$$J_r = L_{Nar}(-\mathscr{F}\Delta\psi) + L_r(-\Delta F). \tag{2}$$

Here $\mathscr{F}$ is the Faraday constant and the L's are phenomenological coefficients. In assigning the same phenomenological cross-coefficient L_{Nar} to each flow we are assuming the validity of the Onsager reciprocal relation, which has been widely tested in a variety of systems [6:15].

If the above formalism is valid, and if the phenomenological coefficients and the free energy ΔF do not change significantly when we vary $\Delta\psi$ briefly, we should find linear relationships between J_{Na}, J_r, and $\Delta\psi$. This has been found to be the case. Studies in the toad bladder have demonstrated a linear relationship between J_{Na} and $\Delta\psi$ over a range of ±100mV[7]. Studies in the frog skin have demonstrated a linear relationship between the rate of oxygen consumption J_r and $\Delta\psi$ over a range of ±160mV [8:77].

These results suggest strongly that the L's and ΔF are not significantly affected by rather extensive perturbations of $\Delta\psi$. This being the case, the above equations may be used to estimate the free energy of the metabolic driving reaction. As you can readily check,

$$\Delta F = \frac{(J_{Na})_{\Delta\psi=0}}{(\partial J_r / \partial \mathcal{F} \, \Delta\psi)_{\Delta F}} \; . \tag{3}$$

The ability to analyze the energetics of active sodium transport simply and precisely by this means has interest in its own right, but equation (3) takes on additional significance in terms of considerations discussed earlier. As you may well imagine, much effort has been expended in attempts to determine the mechanism whereby aldosterone enhances active sodium transport. Some workers believe that aldosterone increases the permeability of a barrier to sodium movement. Others suggest that aldosterone may directly stimulate the "sodium pump", or increase the rate of synthesis of the "energy donor" [4:1185][9:177]. If this latter mechanism were operative it would be expected to increase the negative free energy of the metabolic driving reaction.

We have approached this problem by studies in paired skins from a single frog, one serving as a control, the other being treated with aldosterone. The results indicate that aldosterone does in fact cause a marked increase of the negative free energy of the reaction which drives aerobic sodium transport [10]. It is to be anticipated that these results, obtained by means of concepts more familiar to physicists and physical chemists than to biologists, will lead to meaningful experiments and interpretations in terms of the more customary approaches of biological chemistry.

Finally, a few words about another aspect of kidney function which

we considered earlier, water permeability, and its response to anti-
diuretic hormone.   Here, too, the frog skin and toad bladder serve as
useful model systems.   Both of these tissues permit osmotic water
flow at a rate proportional to the difference in chemical potential
across the membrane.   In both cases the permeability to water is
exquisitely sensitive to ADH; minute amounts of ADH may increase
the rate of water flow by some 10 to 100-fold.   Again, much attention
has been directed at the mechanism of this response.   Early studies
were interpreted as indicating that ADH increases water permeability
by increasing the caliber of the channels for water flow [11][12:60]
[13:933].   More recently Hays has suggested that ADH acts by increasing the number of channels available for water flow [14:1016].

Both of these interpretations have necessarily been based on indirect
experimental evidence, since no means exist for observing the water
channels directly.   Electron microscopy permits fine spatial resolution, but available techniques require fixation of tissues by means
which may markedly alter their state.   Obviously it would be desirable
to utilize a technique which might permit an estimation of the number
and caliber of water channels both in the absence and presence of ADH,
under conditions as near the physiological as possible.   Accordingly, it
is tempting to speculate that one of the techniques described by
Professor King in this book might one day be adapted for this purpose.
Professor King suggests that it might be possible to characterize a
functioning biological tissue by mapping the evaporation of various

molecules from its surface. Such a characterization should provide useful information concerning the mode of action of ADH. We would expect that for quite a while yet the technique would fail to provide the spatial resolution necessary to differentiate between increased numbers and increased size of water channels, but refinements incorporating features of scanning electron microscopy might eventually solve even this problem.

Obviously, what I have just said is highly speculative and much work remains to be done before we can determine the practicality of applying Professor King's elegant technique to this important problem. Whether or not this approach proves useful, it serves to stress the fundamental point of view that underlies all of the work which you are hearing about this week, our belief that the concepts and techniques of the physical sciences can provide important insights into fundamental biological problems which could not easily be obtained by the more traditional approaches.

REFERENCES

[1] Pitts, R. F. 1968. Physiology of the Kidney and Body Fluids. Chicago: Year Book Medical Publishers, Inc.

[2] Ussing, H. H. and K. Zerahn. 1951. Acta Physiol. Scand. 23: 110.

[3] Leaf, A., J. Anderson, and L. B. Page. 1958. J. Gen. Physiol. 41:657.

[4] Sharp, G. W. G. and A. Leaf. 1964. Nature. 202:1185.

[5]   Essig, A. and S. R. Caplan. 1968. Biophys. J. 8:1434.

[6]   Miller, D. G. 1960. Chem. Rev. 60:15.

[7]   Saito, T., P. D. Lief, and A. Essig. Submitted for publication.

[8]   Vieira, F. L.,S. R. Caplan, and A. Essig. 1972. J. Gen. Physiol. 59:77.

[9]   Fanestil, D. D., T. S. Herman, G. M. Fimognari, and I. S. Edelman. 1968. In Regulatory Functions of Biological Membranes, ed. J. Jarnefelt, p. 177. Amsterdam: Elsevier Publishing Company.

[10]   Saito, T., A. Essig, and S. R. Caplan. Unpublished observations.

[11]   Ussing, H. H. 1960. The Alkali Metal Ions in Biology. Berlin: Springer-Verlag.

[12]   Koefoed-Johnsen, V. and H. H. Ussing. 1953. Acta Phsiol. Scand. 28:60.

[13]   Hays, R. M. and A. Leaf. 1962. J. Gen. Physiol. 45:933.

[14]   Hays, R. M., N. Franki, and R. Soberman. 1971. J. Clin. Invest. 50:1016.

CHAPTER 3

PHYSICAL CHEMISTRY OF COLLAGEN IN THE SOLID STATE

Ioannis V. Yannas

INTRODUCTION

Collagen molecules seem to perform in the soft connective tissue of animals somewhat the same function as cellulose molecules perform in plants: aggregated in fibrillar crystallites, they support mechanical stresses which are transferred to them by a low-modulus matrix. Both in healthy and in diseased connective tissue, collagen functions in the solid state. As a component of materials for surgical applications, the performance of collagen also depends largely on its solid state properties. In addition, collagen is beginning to emerge as a material of engineering importance, with industral applications which also demand that this protein perform as a bulk substance. The purpose of this chapter is to review the collagenous solid state, both native and reconstituted, in a way that can make it more accessible for study by physical chemists, biochemists, physicians or materials scientists.

MOLECULAR STRUCTURE

Collagen is a major protein constituent of connective tissue in the vertebrate as well as invertebrate animal kingdoms. In these sources, collagen is often present in the form of macroscopic, sometimes beautifully translucent, fibers which can, with varying degree of ease, be separated mechanically from the host tissue. Molecular collagen can be extracted from such fibers using a variety [1,2] of aqueous solvents and can subsequently be precipitated into several crystalline forms. Collagen from a large number of animal sources has been assidously studied, both in the state of dilute aqueous solution as well as in the crystalline state, in native or reconstituted form, and there is wide agreement [3] that two sets of characteristics can conclusively differentiate it from other proteins: First, the amino acid composition (Table 1) is distinctive in its very high content of glycyl residues (about 33 percent of the total) and of imino-acid prolyl and hydroxy-prolyl residues (about 25 percent of the total). Second, the wide-angle X-ray diffraction pattern of collagen fibers, shows a strong meridional arc corresponding to a spacing of about 2.9 Å and a strong equatorial spot which corresponds, in moist collagen, to a spacing of about 15 Å. No other protein combines these structural features; consequently, their presence in an unknown specimen constitutes sufficient (though not necessary) evidence that the investigator is dealing with some form of collagen. A third, less used but quite distinctive, set of characteristics is the infrared spectrum obtained with fibers or films of collagen.

The structural complexity attainable by a polymer seems to be related to the diversity of bonds which hold together the constituent atoms. This qualitative molecular rule was recognized by Bernal [4]

some time ago who applied the term _heterodesmisity_ to describe the
diversity of bonding interactions in a macromolecule. It predicts, as
an example, that isolated chains of simple hydrocarbon homopolymers
such as polyethylene and polystyrene achieve a smaller number of
conformations and aggregate with other chains in a much smaller
number of ways than the chains of an 18-amino-acid polymer such as
collagen do. It is, in fact, the wealth of types of intra- and inter-
molecular bonds in proteins which clearly sets these macromolecules
apart from even the most complicated synthetic polymers commonly
available today. These considerations account for the fact that a
structural hierarchy, quite unlike anything that has been observed
with synthetic polymers, characterizes the various collagenous states
of matter (Figures 1-4). Elements of this hierarchy, which is largely,
but by no means wholly, understood, will now be reviewed briefly.

Primary structure denotes the complete sequence of amino acids in
a protein. Collagen is made up of 18 amino acids in relative amounts

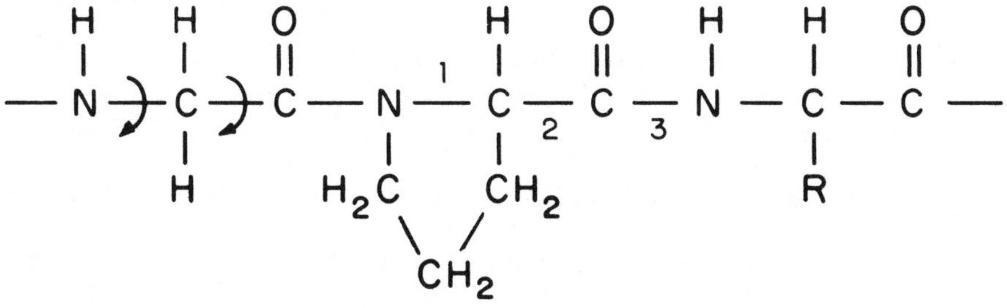

Figure 1. The sequence Gly-Pro-Z. Rotation occurs freely at the
glycyl residue (arrows) unless prevented by chain-chain interactions.
By contrast, rotation is hindered at the site of the prolyl residue
(bonds 1, 2 and 3). Reproduced from I. V. Yannas, Rev. Macromol.
Chem. C7(1), 49 (1972).

A. TYPICAL SEQUENCE

GLY-X-Y-GLY-PRO-Y-GLY-X-HYP-GLY-PRO-HYP

B. MINOR HELIX

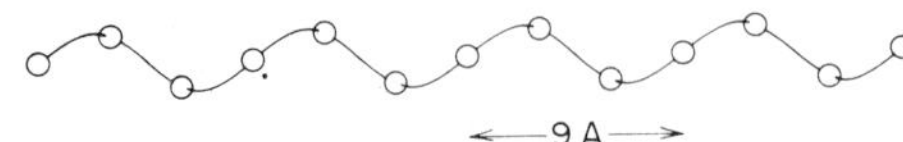

C. MAJOR (TRIPLE) HELIX

D. MOLECULE

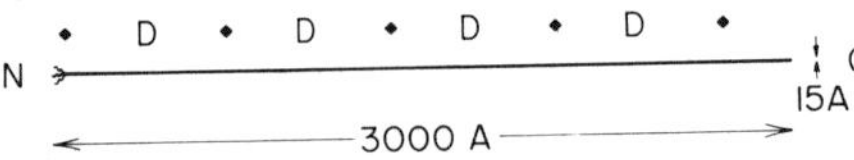

E. FIBRIL

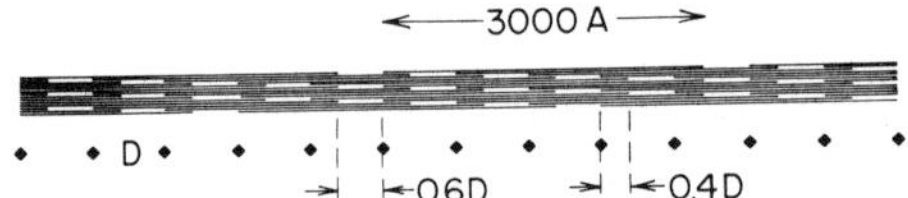

Figure 2.   Diagrammatic representation of several levels of order in collagen: A, primary structure; B, secondary structure; C and D, tertiary structure; E, quaternary structure (details in text). A prominent feature of the tertiary structure is a repeat distance D, about 680Å; each tropocollagen molecule extends 4.4D and a space or "hole" of 0.6 D is left between molecules in line.   Reprinted from Ref. [23].

(amino acid composition) which are well known for several animal species, but in sequences which are still largely undetermined for any one species.   Even though certain differences in amino acid composition do exist between collagens from species widely removed from each other in the evolutionary scale [5] , the composition of collagen from human tendon shown in Table 1 is representative and points at several important facts:

First, collagen is a decidedly hydrophilic protein since the combined content of acidic, basic and hydroxylated amino acid residues exceeds by far the content of lipophilic residues [5] (Table 1). This feature suggests that collagenous states of matter should either dissolve or swell highly in polar liquids with high solubility parameters; and, in fact, the best solvents for such materials known today are aqueous solutions of weak or strong electrolytes as well as highly hydroxylated compounds, such as glycols. In tissues, collagen fibers function in a highly swollen state, often containing up to 60-65 wt. -percent water.

Second, glycyl residues make up one-third of the total (Table 1). Taken together with the additional finding that these residues are spaced evenly almost throughout the chain structure [6] , this fact leads to the expectation that the collagen chains, possessing as they do such a profusion of evenly spaced "hinges" around which relatively unrestricted rotation can occur, should be quite flexible (Figure 1; Figure 2A). This expectation is fulfilled as long as restrictions to rotation around skeletal bonds, arising from chain-chain bonding, are absent. Such restrictions are apparently absent only when the chains of collagen are separated effectively from each other by diluent; this condition materializes with gelatin rubbers, the high extensibility of which demonstrates the intrinsic flexibility of the individual collagen chains. On the other hand, native collagen molecules show evidence of being highly inflexible , a finding which is consistent with their triple-standed character and the severe restrictions to rotation of

Table 1   The Amino Acid Composition of Collagen (Human Tendon)

Trivial name	R in repeating sequence $-NHCHRCO-$; or imino acid formula	Number of amino acid residues per 1000 total residues [5]
Alanine	$-CH_3$	110.7
Glycine	$-H$	324
Valine	$-CH-CH_3$ $\quad\ \ \vert$ $\quad\ CH_3$	25.4
Leucine	$-CH_2-CH\begin{smallmatrix}\diagup CH_3\\[2pt]\diagdown CH_3\end{smallmatrix}$	26.0
Isoleucine	$-CH-CH_2-CH_3$ $\quad\ \vert$ $\quad\ CH_3$	11.1
Proline	$CH_2\!-\!CH_2$ $\ \vert\qquad\vert$ $CH_2\ \ CH-COOH$ $\quad\diagdown\ \diagup$ $\qquad NH$	126.4
Phenylalanine	$-CH_2-\langle\ \rangle$	14.2
Tyrosine	$-CH_2-\langle\ \rangle-OH$	3.6
Serine	$-CH_2-OH$	36.9
Threonine	$-CH-OH$ $\quad\ \vert$ $\quad\ CH_3$	18.5
Methionine	$-CH_2-CH_2-SCH_3$	5.7
Arginine	$-CH_2-CH_2-CH_2-NH-C\begin{smallmatrix}\diagup NH_2\\[2pt]\diagdown NH\end{smallmatrix}$	49.0
Histidine	$-CH_2-C-N$ $\qquad\ \ \Vert\ \ \Vert$ $\qquad HC\ \ CH$ $\qquad\ \ \diagdown\diagup$ $\qquad\quad NH$	5.4
Lysine	$-CH_2-CH_2-CH_2-CH_2-NH_2$	21.6
Aspartic Acid	$-CH_2-COOH$	48.4
Glutamic acid	$-CH_2-CH_2-COOH$	72.3
Hydroxyproline	$HO-CH-CH_2$ $\qquad\ \vert\qquad\vert$ $\quad\ CH_2\ CH-COOH$ $\qquad\ \diagdown\ \diagup$ $\qquad\quad NH$	92.1
Hydroxylysine	$-CH_2-CH_2-CH-CH_2-NH_2$ $\qquad\qquad\quad\ \vert$ $\qquad\qquad\quad OH$	8.9

individual chains which are imposed by chain-chain bonding.

Third, the imino acids proline and hydroxyproline together constitute almost one-fourth of the residues (Table 1). This fact guarantees that, in almost one out of four residues, rotation around the bond denoted 1 in the diagram of the prolyl residue (Figure 1) will not take place unless destruction of covalent bonds occurs, while the presence of the proline ring will also hinder seriously rotation around the bond denoted 2 (Figure 1). If, finally, the partial double bond character of the peptide bond (denoted 3 in Figure 1) is recognized as a significant barrier to rotation around this bond, it becomes clear that the prolyl and hydroxyprolyl residues (almost one-fourth of the total) act as rigid links in the collagen chain.

A fourth characteristic of the primary structure of collagen emerges from sequencing studies and will be examined in greater detail in the next section. Briefly, collagen chains appear to be made up of block sequences, alternately dominated by imino acids and by α-amino acids.

Evidence for the existence of a few non-peptide (perhaps ester) bonds in the backbone of the collagen chains has been presented [7].

<u>Secondary structure</u> refers to the configuration of individual chains. To a large extent, the configuration of polypeptide chains is dominated by the stereochemical features of the constituent amino acid residues. The latter determine the bond lengths and bond angles of the chain back bone besides conferring the possibility of rotation around certain

skeletal bonds. In polypeptides [8], the option of rotating around

skeletal bonds is usually decided in favor of formation of a large

number of N-H $\cdots$ O hydrogen bonds either intramolecularly (e. g. ,

α-helix) or intermolecularly (e. g. , β-structure). In collagen, the

abundance of prolyl and hydroxyprolyl residues prevents [9] the

formation of α-helical chain segments; instead, it appears that long

sequences of the individual chains of collagen resemble the left-handed

version of the considerably steeper poly-L-proline II helix [9] which

has a pitch of about 9 Å (Figure 2B), compared to a pitch of about

5. 4Å for the α-helix. On the other hand, the occurrence of a glycyl

residue at almost every third position appears also to confer to the

individual strands of collagen the capacity of polyglycine II chains for

extensive sterically permissible hydrogen bonding [9] . Individually

considered, therefore, the chains of collagen are helical macromole-

cules with a strong propensity [10] for interchain, rather than

intrachain, hydrogen bonding.

Tertiary structure refers to large-scale folding, helicity and other

convoluted forms assumed in space by the totality of polymeric chains

which constitute the fundamental molecular unit. In collagen, the

fundamental unit is termed the tropocollagen (TC) molecule [11] , a

right-hand superhelix or coiled coil with a repeat distance of about

86 Å , made up of three left-hand strands [9, 12-16] (Figure 2C), each

strand partly resembling in form the poly-L-proline II helix while the

three of them aremutually supported by interchain N-H $\cdots$ O hydrogen bonds [9, 12-16]. The dimensions of tropocollagen molecules have been determined in dilute solution [17] where these units behave as rods about 2800 Å long and about 15 Å in diameter, with a molecular weight which shows some variation according to species and method of extraction but which is found to be very close to 300,000 for most preparations studied so far [1, 9, 17] (Figure 2D). Each of the three polypeptide chains, known as α-chains, which constitute tropocollagen, extends the entire length of the molecule [18] : an α-chain has a molecular weight of about 100,000 and contains, therefore, roughly 1,000 amino acid residues. Of the three chains, two have an almost identical amino acid composition [1, 20] in most preparations studied and are known as α1 chains; the third has a somewhat different composition and is known as the α2 chain. Two α-chains can "dimerize" by covalent crosslinking to form a so-called β-chain while three can form a Y-chain. One of the ends of the tropocollagen molecule, involving 10-20 amino acid residues at the N-terminal ends of each of the α-chains (Figure 2D), is believed to be nonhelical [1, 20-23] and to contain the sites (apparently the lysyl residues) where the covalent crosslinking, responsible for biosynthetic conversion of α-chains to β- and Y- chains in the native state, is believed to occur [1,18-26]. The three strands of the helix are therefore tied together by primary as well as secondary valence bonds, even though the former probably do not make a substantial contribution to the stability of the isolated helix.

STRUCTURE OF THE SOLID STATE

Above the level of tertiary structure, collagen forms quasi-lattices and behaves as a solid substance. The methodology for the study of quaternary and higher structures changes accordingly; rather than using techniques such as ultracentrifugation, viscometry, light scattering and flow birefringence which prove so useful in the study of dilute polymer solutions, investigators of the solid state of collagen employ transmission and scanning electron microscopy, wide- and small-angle X-ray diffraction, infrared spectroscopy, differential thermal analysis and several other techniques.

Quaternary structure in collagen will refer here to small orderly aggregates of tropocollagen molecules. In most synthetic polymers, order at this level (often called the supramolecular level) is largely limited to the folded-chain lamella. In collagen, by contrast, the variety of forces of different strength which hold together atoms within the same chain or in different chains (Bernal's concept of hetero-desmisity [4]) probably accounts for the fact that tropocollagen molecules aggregate to form complex crystallites of several kinds including native crystallites, segment long spacing or SLS crystallites, fibrous long spacing or FLS crystallites, as well as other variants [27-31]. These crystallites can be highly anisometric (native, FLS) or approximately isometric (SLS) and have diameters ranging widely according to the details of the crystallization process, the latter conceivably occurring by precise lateral overlapping and linear aggrega-

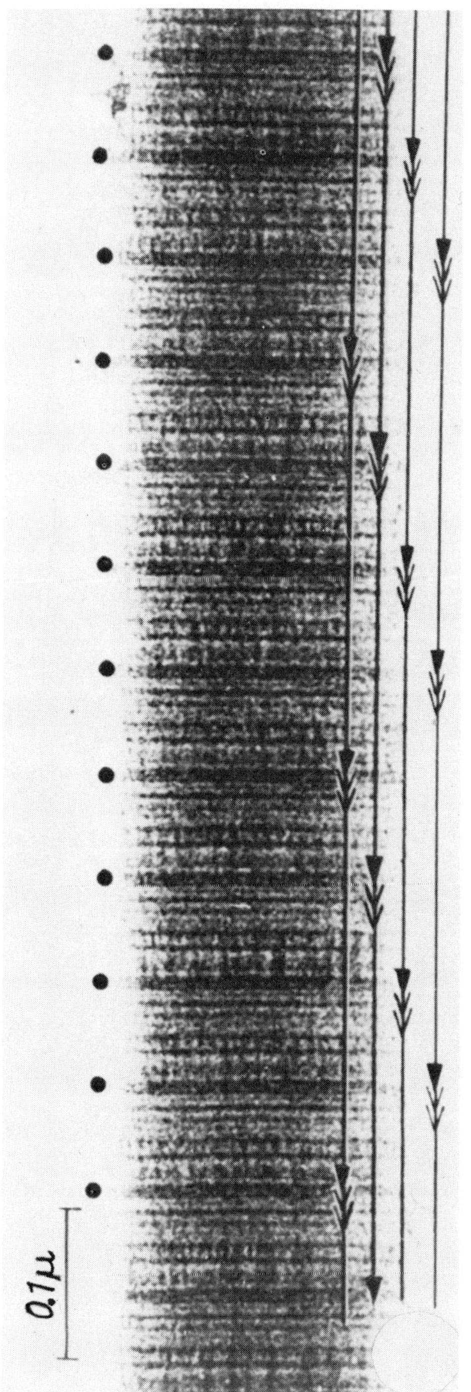

Figure 3.   A fibrillar crystallite showing the periodicity typical of the native form.   The band pattern has a period of about 700Å and a rich intraperiod band pattern is also evident. Tropocollagen molecules are packed with their axes along the fiber direction.   Stained with PTA at pH 4.2.   Reprinted from Ref. [30].

tion of tropocollagen molecules (Figure 3). A pattern of fine cross-striations with an axial period of about 690Å for the native and about 2800Å for the FLS forms characterizes the transmission electron micrographs of stained specimens of these crystallites [27-33]. This pattern (Figure 3), also observed by small-angle X-ray diffraction [34], is known to correspond to a sequence made up of regions (bands) which react with the heavy-metal ionic stains useful in this kind of electron microscopy, alternating regularly with regions that do not stain well (interbands) [31]. Much finer striations (intraperiod bands) have been observed (Figure 3) and evidence for the periodic occurrence of spaces or holes in these crystallites has been presented [31, 35] (Figure 2E). Three-dimensional views of the packing of tropocollagen molecules in such crystallites have been based on models of perfect or defect hexagonally packed lattices [12]. Another more detailed model [2] for the three-dimensional arrangement of tropocollagen molecules has included the fact (well-known from the limited solubility of native collagen fibers) that these molecules are cross-linked and that, in particular, $\alpha 1$ chains are covalently linked to each other, while the $\alpha 2$ chain is covalently linked to an $\alpha 2$ chain in an adjacent triple helix [36]; further consideration of the finding that a tetramer of tropocollagen molecules appears to be the next higher fundamental particle above the level of the single tropocollagen molecule [37] has led to rather complete description of a "limiting micro-fibril" model [2, 38] for the three-dimensional organization of the

Figure 4. Several views of rat tail tendon fibers by scanning electron microscopy. A: fibers treated with 0.5 $\underline{M}$ Na$_2$ HPO$_4$ to show gradual removal of the reticular membrane which covers the fibers. B: fibers treated at one end with 0.05 $\underline{M}$ acetic acid to illustrate details of dissolution process. C: fracture of wet tendon. D: fracture of anhydrous tendon. All specimens coated with gold [39].

collagen molecules in a native fiber. It is quite clear that the order in collagen crystallites clearly surpasses in sophistication that which has been so far attained in any crystalline lattice of man-made polymer chains. Collagen crystallites can be formed either biosynthetically (native forms) or by precipitation from dilute solution (SLS, FLS) but partial recrystallization of collagen from the melt appears to be possible as well.

Considerably less attention has been paid to the analysis of collagen structures at a level higher than the quaternary. A level of order of this kind could be defined in terms of a fundamental particle or repeating unit consisting of an orderly aggregation of small crystallites such as those described in the paragraph immediately above. If such a level of order did not exist, the macroscopic collagen tendon (Figure 4A) could still be considered either as an oriented or as a random array of small collagen crystallites; the latter possibility is, however, contradicted by the X-ray pattern of the tendon where a high level of orientation along the fiber axis is clearly evident. Certain results of a search by scanning electron microscopy for such a higher level of order are shown in Figure 4 [39].

Rat tail tendon fibers, $100-200\mu$ in diameter, have an ellipsoidal cross section and are crimped with a frequency of $100-300\ \mu^{-1}$ (Figure 4A). We treated these fibers with $0.5\underline{M}$ NaH_2PO_4 over a half hour period prior to taking the photograph in Figure 4A in order to remove partly the reticular membrane which covers individual fibers. As a result of such mild treatment, the membrane remained intact on certain fibers and was partly removed exposing the fiber below on others, while a few fibers were completely denuded (Figure 4). The restraining effect of the reticular membrane on the swelling of collagen fibers has been reported by Jordan-Lloyd and Marriott [40] and described in detail by Kwon, Mason and Rigby [41] who also found that the membrane had slightly different properties from the tendon fiber, including a higher shrinkage temperature and a small but significant sulfur content (about 1 per cent on the dry weight).

Details of the dissolution of a denuded tendon fiber in 0.05$\underline{M}$ acetic acid, a well known collagen solvent [1], presented in Figure 4B, where the fiber shown has been treated with solvent at the top end. The collapsing fiber appears to be made up of parallel fibrils, averaging about 10 μ in diameter. We find no evidence in Fig. 4B that the macroscopic tendon fibers are helical as has been suggested by several investigators who based their conclusions on less direct evidence [42-26]. The conclusion that tendon fibers are not helical has been previously arrived at indirectly by Diamant [47].

An interesting perspective of the structural elements of a tendon fiber is provided by viewing the fracture surfaces of these fibers [39]. Tendon fibers were fractured either immediately following removal from the rat tail (when the water content was about 65 percent-wt.), or after dehydrating to various levels. Wet tendon fractured in a highly fibrillar manner and a distinct fracture surface could not be obtained (Figure 4C). The diameters of most of the fractured fibrils were in the range 5-20μ (Figure 4C) while a closer view of the fracture surface of individual fibrils revealed subunits, referred to here as microfibrils, which averaged about 0.1μ in diameter. When fibers were fractured after being allowed to equilibrate at room temperature (water content about 15 percent-wt), the fracture was also highly fibrillar and structural units of the same size as shown in Figure 4C were observed. A significantly "cleaner" fracture surface was obtained by previously dehydrating the tendon fiber to a moisture content less than 0.1 percent-wt [39]. Figure 4D presents a fracture surface obtained in this manner. Clearly shown in Figure 4D are fibrils, approximately circular in cross section and averaging about 10μ in diameter. Study of the longitudinal section of such a fibril at higher magnification shows that the fibrils are bundles made up of microfibrils, the latter averaging about 0.1μ in diameter and arranged in strictly parallel array along the fiber axis. These results support and extend the early observations of Kuntzel [48] who examined the microtomed cross section of tendon fibers by phase-contrast optical microscopy and found them to be subdivided into fibrils having diameters in the range 5-10μ. A magnified view of the cross section in Figure 4D also supports directly several previous observations in tendon fibers of microfibrils having diameters of 0.1-0.2μ; such units were observed by transmission electron microscopy [49,50] and small-angle X-ray diffraction [51]; their presence has been also predicted by calcu-

lations based on stress-strain data [47] . A preliminary conclusion
from these studies by scanning electron microscopy is that a 200-μ
tendon fiber is analogous to a crimped but untwisted ply yarn made up
of 10-μ untwisted multifilament yarns (the dimensions refer to diameters). Each of the 10-μ multifilament yarns appears, in turn, to be
made up of 0. 1-μ filaments.

BRIEF SURVEY OF PHYSICOCHEMICAL PROPERTIES OF THE SOLID STATE

Extensive results from chemical sequencing work, transmission
electron microscopy and small-angle X-ray diffraction are consistent
with the view [52] that collagen chains are made up of subunits rich in
imino acids (Pro, Hypro) alternating with subunits poor in imino acids.
A solid-state model of collagen as a block copolymer can be used as a
useful basis for qualitative discussion and classification of diverse
optical, thermal and mechanical properties of the solid state.

A distinction between the native and the denatured collagen helix in
the solid state can be clearly made by optical rotation [53, 54] and by
viscoelastic measurements [55, 56]. Infrared spectroscopy [55, 57] ,
differential thermal analysis [58-61] and wide-angle X-ray diffraction
[62] can also be used to make such a distinction. The analytical
interpretation of the results from these solid-state methods is currently at varying levels of development.

The collagen helix undergoes only minor distortion even when dehydrated below an apparent moisture level of 0. 1 wt. -percent and recovers fully upon rehydration [63] . However, both collagen [63] and gelatin [64] , the denatured form of collagen, become covalently crosslinked, and therefore insoluble, when dehydrated below ca. 1 wt. - percent moisture.

Several sources agree that dehydrated collagen melts in the range $200-230^{\circ}C$ [39,58,59,61,65]. The extent to which melting is complicated by non-oxidative pyrolysis is currently under study although results with gelatin have shown that the loss in specimen weight up to $230^{\circ}C$ is about 2 percent [60] . X-ray evidence shows that the collagen helix collapses above the melting point but that it is restored by cooling to room temperature and immersing in water [39] .

Amorphous collagen, in the form of gelatin films cast above $35^{\circ}C$, undergoes a major glass transformation in the region $175-200^{\circ}C$ [56, 60, 66-69] . Depending on the temperature, content of diluent and crosslink density, a gelatin specimen can behave as a rigid, brittle solid. as a rubber capable of extending by 700 percent or as a viscous liquid [56] . However, specimens incorporating the intact collagen helix appear intrinsically incapable of behaving as rubbers under equivalent conditions [55] .

Native collagen fibers are aggregates of crimped filaments [70-73] and stresses less than 10^{7} dynes/ cm^{2} suffice to extend these fibers (uncrimping) by several percent. Certain components of the non-collagenous matrix of these fibers can contribute significantly to the mechanical behavior of native fibers [74, 75]. Fracture of wet tendon

apparently occurs by slippage of filaments having diameters in the

range 5-10μ (Figure 4C) while anhydrous tendon fractures without

evidence of such slippage (Figure 4D).

REFERENCES

[1]   Piez, K. A. 1967. In Treatise on Collagen, Vol. 1, Chap. 5,
ed. G. N. Ramachandran. London and New York: Academic Press.

[2]   Veis, A. 1967. In Treatise on Collagen, Vol. 1, Chap. 8, ed.
G. N. Ramachandran. London and New York: Academic Press.

[3]   Harrington, W. F. In Encyclopedia of Polymer Science and
Technology, eds. H. F. Mark, N. G. Gaylord, and N. M. Bikales,
Vol. 4, p. 1. New York: Interscience.

[4]   Bernal, J. D. 1958. Disc. Faraday Soc. 25:7.

[5]   Eastoe, J. E. 1967. In Treatise on Collagen, Vol. 1, Chap. 1,
ed. G. N. Ramachandran. London and New York: Academic Press.

[6]   Hannig, K. and A. Nordwig. 1967. In Treatise on Collagen,
Vol. 1, Chap. 2, ed. G. N. Ramachandran. London and New York:
Academic Press.

[7]   Gallop, P. M., O. O. Blumenfeld and S. Seifter. 1967. In
Treatise on Collagen, Vol. 1, Chap. 7, ed. G. N. Ramachandran.
London and New York: Academic Press.

[8]   Bamford, C. H., A. Elliott and W. E. Hanby. 1956. Synthetic
Polypeptides. New York: Academic Press.

[9]   von Hippel, P. H. 1967. In Treatise on Collagen, Vol. 1,
Chap. 6, ed. G. N. Ramachandran. London and New York: Academic
Press.

[10] Carver, J. P. and E. R. Blout. 1967. In Treatise on Collagen,
Vol. 1, Chap. 9, ed. G. N. Ramachandran. London and New York:
Academic Press.

[11] Schmitt, F. O. , J. Gross and J. H. Highberger. 1955. J. Exp. Cell. Res. , Supplement 3, p. 326.

[12] Ramachandran, G. N. 1967. In Treatise on Collagen, Vol. 1, Chap. 3, ed. G. N. Ramachandran. London and New York: Academic Press.

[13] Ramachandran, G. N. and G. Kartha. 1965. Nature.174:269.

[14] Rich, A. and F. H. C. Crick. 1955. Nature. 176:915.

[15] Rich A. and F. H. C. Crick. 1961. J. Mol. Biol. 3:483.

[16] Rich A. and F. H. C. Crick. 1958. In Recent Advances in Gelatine and Glue Research, ed. G. Stainsby, p. 20. London: Pergamon Press.

[17] Boedtker, H. and P. Doty. 1956. J. Am. Chem. Soc. 78:4267.

[18] Kang, A. H. , Y. Nagai, K. A. Piez and J. Gross. 1966. Biochemistry. 5:509.

[19] Nold, J. G. , A. H. Kang and J. Gross. 1970. Science. 170: 1096.

[20] Schmitt, F. O. 1959. In Connective Tissue, Thrombosis and Atherosclerosis, ed. I. H. Page. New York: Academic Press.

[21] Hodge, A. J. , J. H. Highberger, G. G. J. Deffner and F. O. Schmitt. 1960. Proc. Nat. Acad. Sci. 46:197.

[22] Rubin, A. L. , D. Pfahl, P. T. Speakman, P. F. Davison and F. O. Schmitt. 1963. Science. 139:37.

[23] Piez, K. A. 1969. In Aging of Connective and Skeletal Tissue. Nordiska Bokhandelns Forlag. Stockholm.

[24] Kang, A. H. , P. Bornstein and K. A. Piez. 1966. Fed. Proc. 25:716.

[25] Schmitt, F. O., L. Levine, M. P. Drake, A. L. Rubin, D. Pfahl and P. F. Davison. 1964. Proc. Nat. Acad. Sci. 51:493.

[26] Piez, K. A. 1968. Ann. Rev. Biochem. 27:547.

[27] Highberger, J. H. , J. Gross and F. O. Schmitt. 1950. J. Am. Chem. Soc. 72:3321.

[28] Highberger, J. H. , J. Gross, and F. O. Schmitt. 1951. Proc. Nat. Acad. Sci. 37:286.

[29] Schmitt, F. O. , J. Gross, and J. H. Highberger, 1953. Proc. Nat. Acad. Sci. 39:459.

[30] Hodge, A. J. and F. O. Schmitt. 1960. Proc. Nat. Acad. Sci. 46:186.

[31] Hodge, A. J. 1967. In Treatise on Collagen, Vol. 1, Chap. 4, ed. G. N. Ramachandran. London and New York: Academic Press.

[32] Hall, C. E. , M. A. Jakus, and F. O. Schmitt. 1942. J. Am. Chem. Soc. 64:1234.

[33] Wolpers, C. 1943. Klin. Wschr. 22:624.

[34] Bear, R. S. 1942. J. Am. Chem. Soc. 64:727.

[35] Petruska, J. A. and A. J. Hodge. 1964. Proc. Nat. Acad. Sci. 51:871.

[36] Veis, A. and J. Anesey. 1965. J. Biol. Chem. 240:3899.

[37] Veis, A. , J. Anesey, and J. Cohen. 1962. Arch. Biochem. Biophys. 98:104.

[38] Veis, A. , J. Anesey, and S. Mussell. 1967. Nature. 215:931.

[39] Huang, C., 1971. M.S. Thesis, Department of Mechanical Engineering, MIT.

[40] Jordan-Lloyd, D. and R. H. Marriott. 1935. Proc. Roy. Soc. B118:438.

[41] Kwon, D. S. , P. Mason, and B. J. Rigby. 1964. Nature. 201: 159.

[42] Heringa, G. C. and H. A. Lohr. 1924. Verhand. Konikl. Ned. Akad. Wet. 1129:1081.

[43] Lerch. H. 1949. Morphol. Jahrb. 90:192.

[44] Verzar, F. 1955. Helv. Physiol. Pharmacol. Acta 13:C64.

[45] Banga, I. , J. Balo, and D. Szabo. 1956. Acta Morphol. Acad. Sci. Hung. 6:391.

[46]   Verzar, F. 1964. Int. Rev. Connective Tissue Res. 2:247.

[47]   Diamant, J. 1970. M.S. Thesis, Case Western Reserve University.

[48]   Küntzel, A. 1941. Kolloid-Z. 96:273.

[49]   Schmitt, F. O., E. Hall, and M. A. Jakus. 1942. J. Cell. Comp. Physiol. 20:11.

[50]   Schmitt, F. O. and J. Gross. 1948. J. Am. Leather Chem. Assoc. 43:658.

[51]   Bolduan, O. E. A. and R. S. Bear. 1950. J. Polymer Sci. 6:271.

[52]   Bear, R. S. 1952. Adv. Protein Chem. 7:69.

[53]   Robinson, C. and M. J. Bott. 1951. Nature. 168:325.

[54]   Yannas, I. V., C. Huang and N-H. Sung, unpublished data.

[55]   Yannas, I. V. and C. Huang. 1972. Macromolecules. 5:99.

[56]   Yannas, I. V. 1968. J. Polymer Sci. A2. 6:687.

[57]   Bradbury, E. M., R. E. Burge, J. T. Randall, and G. R. Wilkinson. 1958. Disc. Faraday Soc. 25:173.

[58]   Okamoto, Y. and K. Saeki. 1964. Kolloid-Z. 194:124.

[59]   Puett, D. 1967. Biopolymers. 5:327.

[60]   Yannas, I. V. and A. V. Tobolsky. 1968. European Polymer J. 4:257.

[61]   Jolley, J. E. 1970. Photo Sci. Eng. 14:169.

[62]   Bradbury, E. M. and C. Martin. 1952. Proc. Roy. Soc. A214: 183.

[63]   Yannas, I. V. and N-H. Sung, unpublished data.

[64]   Yannas, I. V. and A. V. Tobolsky. 1967. Nature. 215:509.

[65]   Oth, J. F. M. 1959. Kolloid-Z. 162:124.

[66] Koleske, T. V. and T. A. Faucher. 1965. *J. Phys. Chem.* 69: 4040.

[67] Yannas, I. V. and A. V. Tobolsky. 1964. *J. Phys. Chem.* 68: 3880.

[68] Gillham, J. K. 1966. *Applied Polymer Symposia* No. 2. 45.

[69] Yannas, I. V. and A. V. Tobolsky. 1966. *J. Macromol. Chem.* 1:723.

[70] Rigby, B. J., N. Hirai, J. D. Spikes and H. Eyring. 1959. *J. Gen. Physiol.* 43:265.

[71] Verzar, F. 1957. *Gerontologia.* 1:363.

[72] Abrahams, M. 1967. *Med. Biol. Engng.* 5:433.

[73] Diamant, J. A. Keller, M. Litt, E. Baer, and G. Arridge, personal communication.

[74] Shields, G. S. 1948. B.S. Thesis, MIT.

[75] Partington, F. R. and G. C. Wood. 1963. *Biochem. Biophys. Acta.* 69:485.

CHAPTER 4

COOPERATIVITY IN BIOLOGICAL SYSTEMS:
A General Quantitative Model with Specific Application to Hemoglobin Oxygenation

Judith Herzfeld and H. Eugene Stanley

PHYSIOLOGICAL SIGNIFICANCE OF COOPERATIVITY

Cooperativity plays a very important role in the regulation of biologi-
cal activity. Let us consider the enzyme catalyzed step, $\alpha \xrightarrow{E} \beta$, of
the metabolic pathway shown in Fig. 1. Production of the product β
is controlled by the binding of substrate α to the catalytic sites of the
enzyme. Three types of substrate binding curves are shown in
Fig. 2a. When binding to different catalytic sites is independent, the

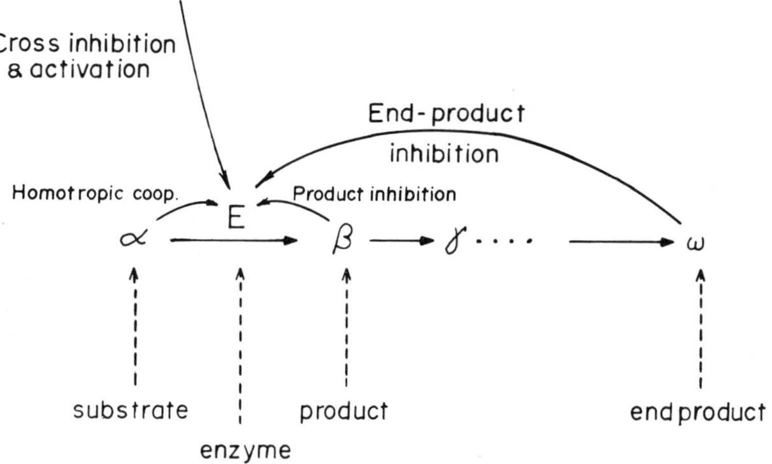

Figure 1.   A schematic representation of a metabolic pathway with
enzyme catalysis (E) and feedback due to product inhibition by the
product β , end-product inhibition by the end-product ω , and cross
inhibition and activation by the metabolites of other pathways.

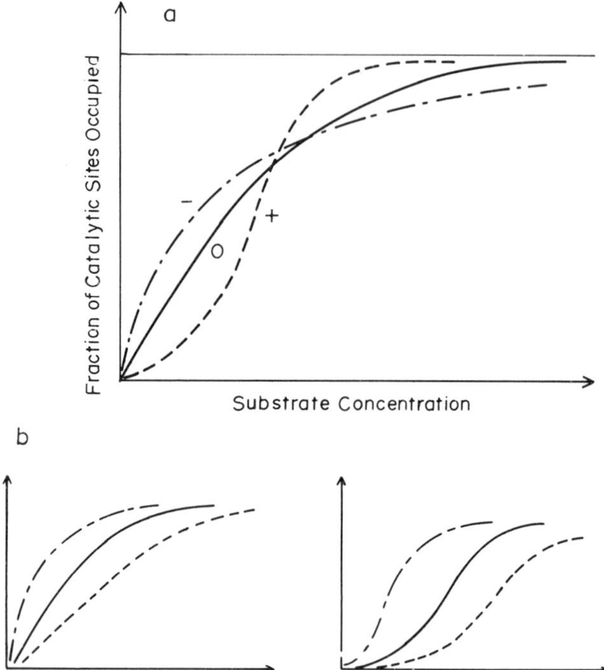

Figure 2.   (a) Homotropic cooperativity: substrate binding curves for independent binding (0), and for positive (+) and negative (−) cooperativity. (b) Heterotropic cooperativity: shift of the substrate binding curve in the presence of inhibitor (i) and activator (a), for two kinds of homotropic interactions.

fraction of sites occupied by substrate is linear with substrate concentration at low concentrations, and levels off as the sites start to fill up. However, if substrate binding at some sites inhibits subsequent binding at other sites (negative homotropic cooperativity), then the slope of the curve decreases much faster. Similarly, if substrate binding at some sites facilitates subsequent binding at other sites (positive homotropic cooperativity), then the slope of the curve increases before it finally levels off. This sigmoid, or S-shape curve is characteristic of positive homotropic cooperativity.

Substrate binding is also influenced by the binding of other metabolites, called effectors. Some effectors are substrate analogs and inhibit catalysis by competing with substrate for binding to the catalytic sites (e.g. product inhibition). Other effectors bind to noncatalytic sites on the enzyme molecule, thereby either inhibiting or facilitating substrate binding at the catalytic sites (e.g. end-product inhibition, cross activation). The influence of effector on substrate binding, called heterotropic cooperativity, is illustrated in Fig. 2b.

The importance of cooperativity for enzyme function is illustrated by the dose-response curves in Fig. 3. We see that the catalysis rate (or response) is more sensitive to changes in substrate concentration (or dose) if the catalytic sites are positively cooperative (Fig. 3b), rather than independent (Fig. 3a). At a single substrate concentration, the catalysis rate is more sensitive to changes in effector concentration, if the catalytic sites are positively cooperative (Fig. 3d), rather than independent (Fig. 3c).

HEMOGLOBIN OXYGENATION AND EFFECTORS OF HEMOGLOBIN OXYGENATION

Cooperativity is also observed in proteins other than enzymes. For example, oxygen binding to the four heme sites of hemoglobin exhibits positive homotropic cooperativity. Positive cooperativity is more conveniently studied in hemoglobin than in enzymes because

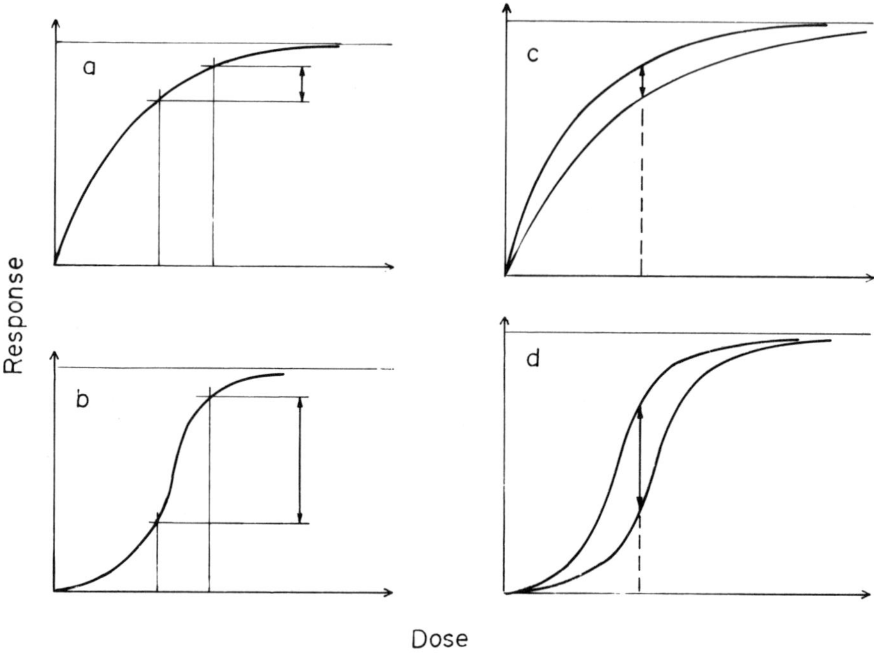

Figure 3.   Sensitivity of enzyme response to substrate and effector concentrations.   Change of enzyme response with change of substrate dosage for independent (a) and positively cooperative (b) sites. Change of enzyme response with change of effector concentration for independent (c) and positively cooperative (d) sites.

substrate binding is measured directly rather than being inferred from a measured catalysis rate.

The significance of cooperativity for hemoglobin is illustrated in Figs. 4-6.   Hemoglobin binds oxygen in the lungs and releases it in the capillaries where some of it is bound by myoglobin and stored in the tissues.   We see, in Fig. 4, that the efficiency of oxygen transport by hemoglobin (Hb) depends on the steepness of the oxygenation curve between the partial pressures of oxygen in the lungs and in the capillaries.   Myoglobin (Mb) has only one oxygen binding heme site

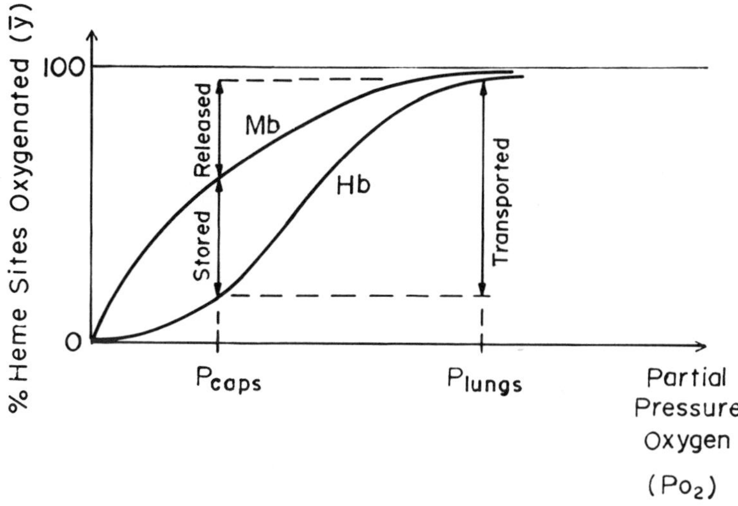

Figure 4. Transport of oxygen by hemoglobin (Hb) and storage of oxygen by myoglobin (Mb).

per molecule, and does not exhibit cooperativity.

One of the effectors of hemoglobin oxygenation is the hydrogen ion. In the physiological pH range, a decrease in pH causes a decrease in oxygen binding as shown in Fig. 5. We see that as low-waste, highly-oxygenated blood travels from the lungs to the capillaries, oxygen is released due to both the decrease in the oxygen pressure and the decrease in pH. By this mechanism the tissues with a high metabolism rate, and a correspondingly high production of CO_2, receive the required extra oxygen.

Oxygenation is also sensitive to the concentrations of certain salts. In particular, the inhibition of oxygenation by 2, 3-diphosphoglycerate (DPG), as illustrated in Fig. 6, is important for adjustment to

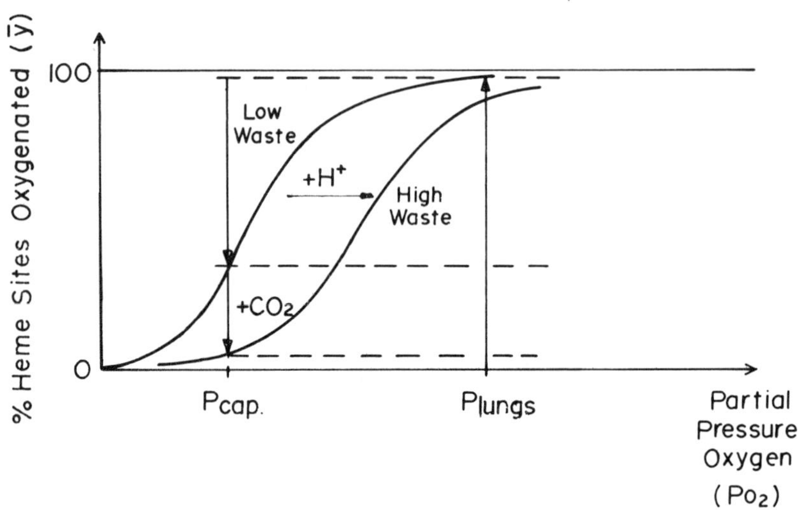

Figure 5.   The relationship between the disposal of CO_2 and the delivery of oxygen in the circulatory system.

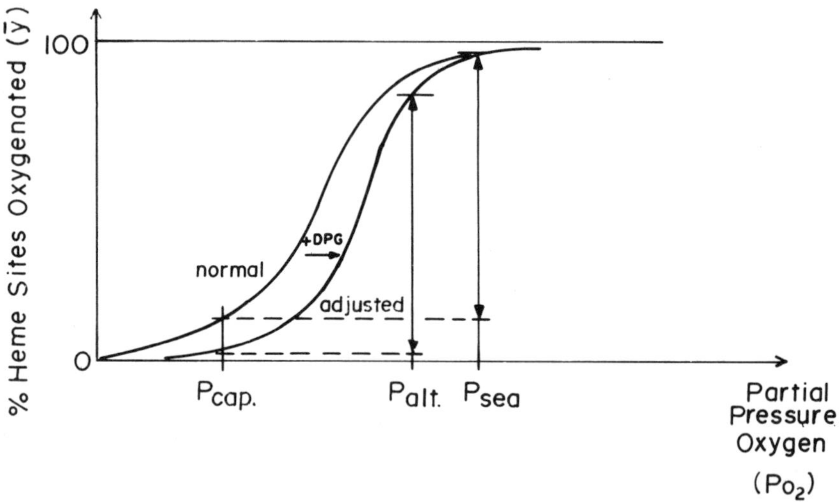

Figure 6.   High altitude adjustment by DPG inhibition.

high altitudes. The body's immediate response to low atmospheric

oxygen pressures is to increase its cardiac output; in the long-term,

it compensates by increasing the hemoglobin concentration. However,
during the intervening time, the body adjusts to high altitude by increasing the red cell DPG concentration so that lower oxygenation in
the lungs is compensated for by extra unloading in the capillaries
[1:187].

Although a number of quantitative models have been proposed to
explain cooperative phenomena, none of these properly represents
the cooperative mechanism of hemoglobin deduced by Perutz on the
basis of recent experiments [2:726]. Moreover, each of these theories describes only a single oxygenation curve at a time. We have
been developing a general microscopic model of cooperativity [3]
which is considerably more versatile in that it can describe any
number of oxygenation curves, taken at various effector concentrations (i. e. describe both homotropic and heterotropic cooperativity),
with a single equation.

PREVIOUS TREATMENTS OF COOPERATIVITY

In 1935, Pauling [4:186] showed that homotropic cooperativity in
hemoglobin could be explained by direct interactions between oxygen
molecules bound to different heme sites. It has since been found that
the distances between the four heme sites is too large for such direct
interactions to account for the strength of the observed cooperativity.
Thus it came to be appreciated that there must be some indirect
interaction mediated by the structure of the protein molecule (allo-

steric interaction). Two such modes of interaction are shown schematically in Fig. 7 (using the conformational terminology defined in
Fig. 8). In both cases substrate binding affects, and is affected by,
the tertiary structure of the subunit involved. In the tertiary-tertiary
("t-t") mode there is an interaction between neighboring subunits; in
the tertiary-quaternary-tertiary ("t-q-t") mode, all the subunits
constrain, and are constrained by, the quaternary structure.

Fig. 9 illustrates the use of these concepts of allosteric interactions in various models of homotropic cooperativity for a tetramer
(e. g. hemoglobin). In 1965, Monod, Wyman and Changeux (MWC)
[5:88] proposed that a cooperative protein molecule has only two
possible structures (each of which is symmetric in that each subunit
is equivalently situated within the quaternary structure and constrained to the same tertiary conformation) and that substrate binds preferentially to one of these two structures. When substrate binds, the

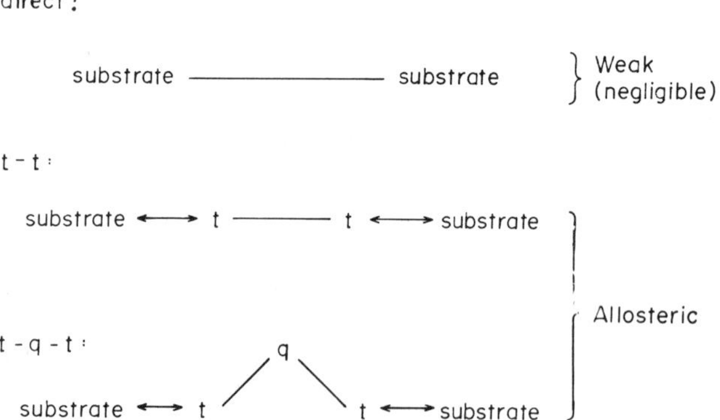

Figure 7.   Schematic representation of different modes of homotropic
interaction. 't' represents the tertiary conformation of a subunit and
'q' represents the quaternary conformation of the molecule.

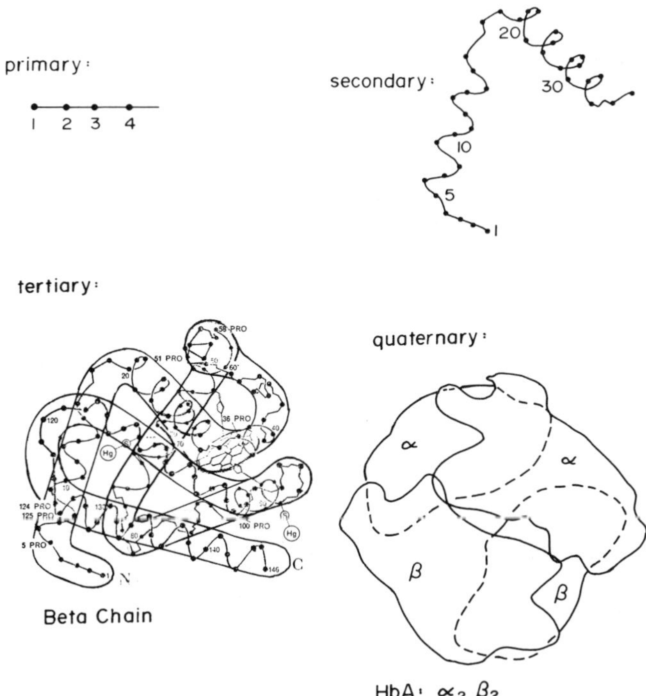

Figure 8.   Specification of protein structure: primary structure - the sequence of amino acids; secondary structure - the structure of the chain of amino acid residues (e. g. α -helix, random coil); tertiary structure - the folding of the secondary structure into a compact unit; quaternary structure - the arrangement of the tertiary units relative to one another in the molecule.

preferred state of the molecule is stabilized relative to the other state, thereby shifting the conformational equilibrium toward the state with the higher substrate affinity and facilitating further substrate binding. This model thus can explain positive, but not negative, homotropic cooperativity.

A second model proposed by Koshland, Nemethy, and Filmer [6: 365], can explain both positive and negative homotropic cooperativity.

Model	Binding	Allosteric Mechanisms	Schematic Diagram
General	Preferential binding	Tertiary-tertiary interactions / Quaternary constraints	
Monod, Wyman, and Changeux [5]	Preferential binding	No tertiray-tertiary interactions / Infinite quaternary constraints (symmetry)	
Koshland, Nemethy, and Filmer [6]	Induced-fit binding (Infinitely preferential binding)	Tertiary-tertiary interactions / No quaternary constraints	
Perutz [2]	Induced-fit binding (Infinitely preferential binding)	No tertiary-tertiary interactions / Quaternary constraints	

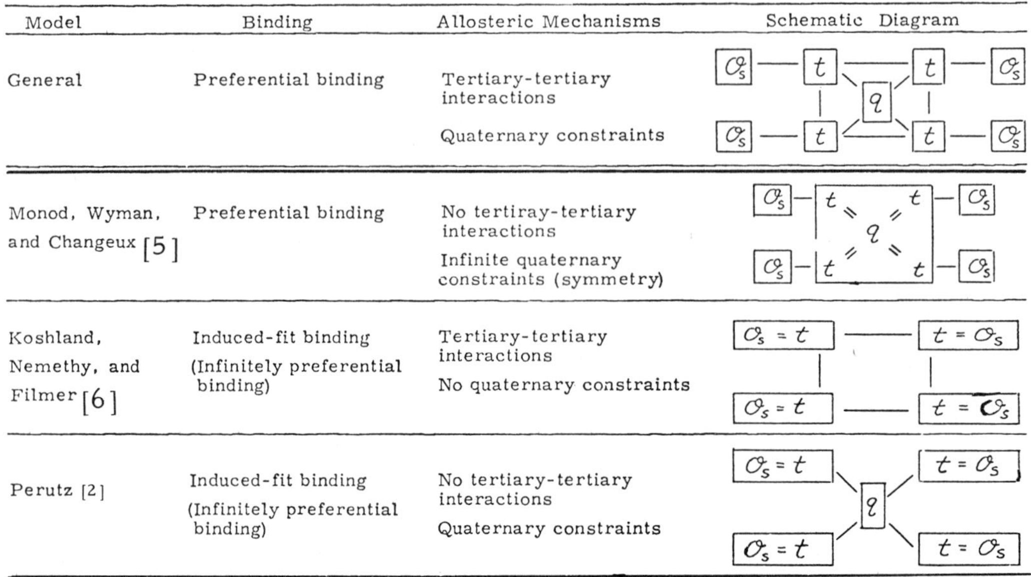

Figure 9. Various allosteric mechanisms for homotropic cooperativity in a tetramer are shown as special cases of the general model. The last column indicates schematically this relation. The notation $\mathcal{O}_s$, t, q corresponds respectively to substrate binding, tertiary conformation, and quaternary conformation. A solid line indicates an interaction, while a double line, together with an enclosing box, indicates an infinitely strong interaction.

In this model substrate binding induces a change in the subunit tertiary structure (called induced-fit binding) and either positive or negative interactions occur between neighboring subunits in the same tertiary conformation. This model has the drawback, relative to the MWC model, that it only explains cooperativity by assuming that substrate binding and conformational change are simultaneous.

Our general model combines all the concepts of the above two earlier theories in order to provide a unified framework for discussion of cooperativity, and to derive a single quantitative description from which to specialize for specific cases. In particular, one limiting case of our general model describes the mechanism of hemoglobin oxygenation proposed by Perutz, which combines the induced-fit binding concept of the Koshland theory and the quaternary constraint concept of the MWC theory [2:726].

QUANTITATIVE DESCRIPTION OF HOMOTROPIC COOPERATIVITY

In order to make any comparisons with data it is necessary to translate the pictures of Fig. 9 into quantitative expressions for substrate binding. The fraction, $\overline{Y}$, of substrate sites occupied may be expressed in the form

$$\overline{Y} = \sum_{\substack{\text{all} \\ \text{states}}} (n_{\text{state}}/N)\, f_{\text{state}} \;, \tag{1}$$

where the summation is over all possible states of the protein molecule, n_{state} is the number of substrate sites occupied in a particular state, N is the total number of substrate sites for a single protein molecule, and f_{state} is the fraction of protein molecules in the system which are in a particular state. We make the statistical assumption that, for a large number of molecules, the fraction which is in a given state is equal to the probability that a single molecule is in that state:

$$f_{\text{state}} = P(\text{state}) \;. \tag{2}$$

This probability is given by the distribution

$$P(\text{state}) = Z^{-1} \exp(n_{\text{state}} \mu/kT) \exp(-E(\text{state})/kT) \,, \tag{3}$$

where the normalization factor, called the partition function, is given by

$$Z = \sum_{\substack{\text{all} \\ \text{states}}} \exp(n_{\text{state}} \mu/kT) \exp(-E(\text{state})/kT) \,. \tag{4}$$

Here E(state) represents the energy of a particular state, and the activity of the substrate, $\exp(\mu/kT)$, is assumed to be approximately proportional to the partial pressure of substrate in the gas phase and the free concentration of substrate in solution. Combining Eqs. (1) – (4), we can express the fraction of substrate sites occupied as

$$\overline{Y} = \frac{1}{N} \frac{1}{Z} \frac{\partial Z}{\partial(\mu/kT)} \,. \tag{5}$$

Thus the problem reduces to calculating the partition function, Z. In order to calculate Z we need to know the energies E of all the states of the protein molecule. To do so it is necessary to specify the state of the molecule uniquely. We choose the notation

$$\text{state} = \{\sigma_i; \ t_i; \ q\} \,. \tag{6a}$$

Here σ_i takes on the values 0, 1 according as the ith site is vacant or occupied, $t_i=k$ indicates that the ith subunit is in the kth tertiary conformation, and $q=m$ indicates that the molecule is in the mth quaternary conformation. This treatment is general for any system displaying homotropic cooperativity. For a tetramer (N=4), such as homoglobin, Eq. (6a) becomes

$$\text{state} = \{\sigma_1, \ \sigma_2, \ \sigma_3, \ \sigma_4; \ t_1, t_2, t_3, t_4; \ q\} \,. \tag{6b}$$

For hemoglobin it suffices to assume that there are only two possible quaternary conformations for the molecule (m=1,2) and only two possible tertiary conformations for a subunit (k=1,2). The energy of a particular state of the molecule can then be expressed in terms of its state variables as follows:

$$E(\text{state}) = U_q q + \sum_{i=1}^{N} \{ U_t t_i + (U_\sigma + (-1)^{t_i} U_{\sigma t}) O_i$$

$$-U_{tt}(-1)^{t_i + t_{i+1}} - U_{qt}(-1)^{t_i + q} \} \;, \tag{7}$$

where U_q is the difference in energy between the two quaternary conformations of the molecule; U_t is the difference in energy between the two tertiary conformations of a subunit; $(U_\sigma + (-1)^{t_i} U_{\sigma t})$ is the energy of substrate binding to a subunit in the t_i tertiary conformation; $-U_{tt}(-1)^{t_i + t_{i+1}}$ is the energy of interaction between the tertiary structures of neighboring subunits in the tertiary conformations t_i and t_{i+1}; and $-U_{qt}(-1)^{t_i + q}$ is the energy of the constraint of a subunit in the t_i tertiary conformation when the molecule is in the q quaternary conformation. Eqs. (4), (5), and (6) together describe substrate binding in a constant chemical environment; we must expand this theory in order to describe the effects of changes in effector concentrations.

HETEROTROPIC COOPERATIVITY

Effectors can influence substrate binding by the three basic mechanisms illustrated in Fig. 10. Effectors may interact directly (direct effectors) with the substrate or the substrate binding site, altering the stability of the bound complex. In the case of extreme antagonistic interactions, effector binding excludes substrate binding (competitive effector). Effectors may also bind preferentially to different tertiary conformations (tertiary effectors), thereby shifting the equilibrium between conformations and changing the substrate binding affinity. Similarly, effectors may bind preferentially to different quaternary conformations (quaternary effectors), thereby shifting the equilibrium between conformations and changing the substrate binding affinity.

In the case of hemoglobin oxygenation, carbon monoxide is an example of a competitive effector. Hydrogen ions are tertiray effectors, binding preferentially to the intact salt bridges of the deoxy tertiary conformation. 2, 3-diphosphoglycerate (DPG) is a quaternary effector, binding preferentially to the deoxy quaternary conform-

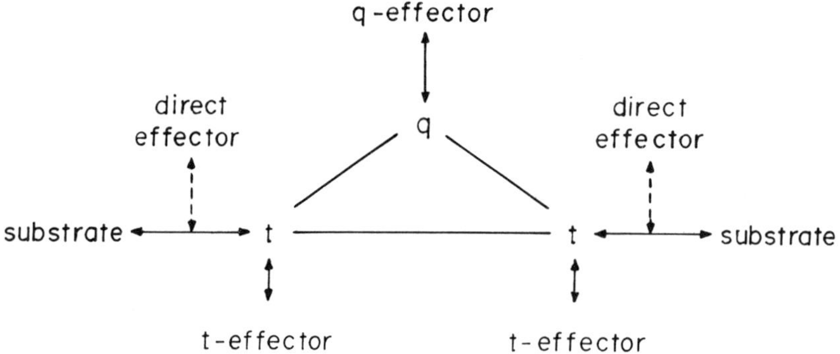

Figure 10.   Schematic representation of different modes of effector action.

ation.

If we assume that DPG is purely a quaternary effector, the energy of DPG binding depends only on the quaternary conformation of the hemoglobin molecule. Thus, in order to describe the effect of DPG on the hemoglobin system, we add to the energy in Eq. (7) the additional DPG binding energy

$$(U_{\tilde{\sigma}} + (-1)^q U_{\tilde{\sigma}q})\,\tilde{\sigma} \quad , \tag{8}$$

where $\tilde{\sigma}$ is the state variable indicating the presence ($\tilde{\sigma}=1$) or absence ($\tilde{\sigma}=0$) of DPG. With the two parameters $U_{\tilde{\sigma}}$ and $U_{\tilde{\sigma}q}$, in addition to the three parameters U_q, U_σ, and U_{qt} of the Perutz system ($U_{tt}=0$ in the absence of t-t interactions and $U_t=\infty$, $U_{\sigma t}=\infty$ for induced fit binding), we can describe hemoglobin oxygenation at many different DPG levels by using Eqs. (4) and (5) and the matrix methods of Thompson [7:110]. This five parameter fit is shown in Fig. 11 under two different sets of pH and chloride conditions.

QUATERNARY STRUCTURE AND OXYGENATION

By fitting oxygenation data we can determine the values of the energy parameters of the system. From these we can calculate the probabilities of each of the states of the system according to Eq. (3). In particular, we can calculate the probability that a hemoglobin molecule with n heme sites oxygenated will be in one or the other of the two quaternary conformations. The results of this calculation are

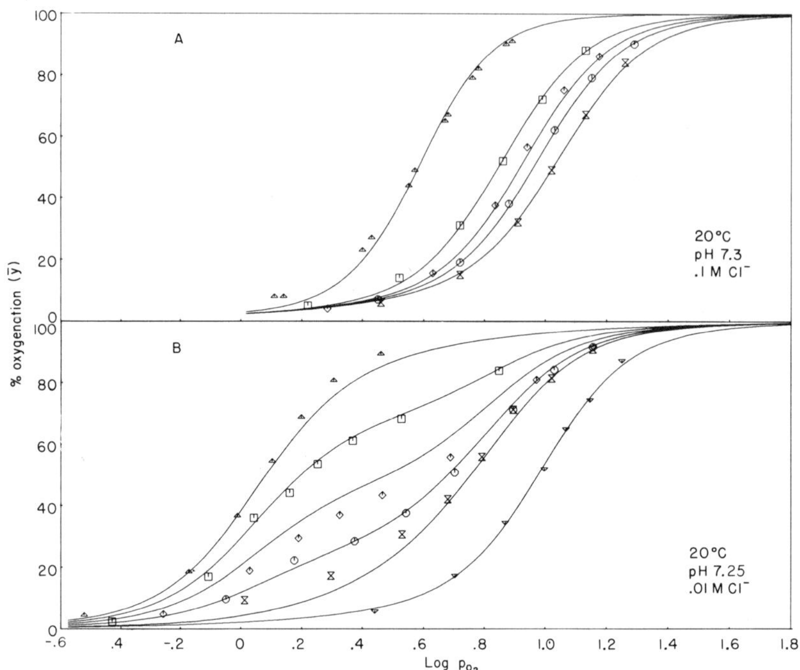

Figure 11.   Experimental oxygenation data compared with the
theoretical equilibrium curves.

A.   Data points [10:2567] taken with 6×10^{-6}M hemoglobin (by tetramer)
and 0.05M BisTris buffer.   Total DPG: △ none; ▢ 1.3×10^{-4} M;
◇ 2.5×10^{-4}M; ○ 4.0×10^{-4} M; ⊠ 1.0×10^{-3} M.
Theoretical equilibrium curves shown for parameter values:
U_q =6.72kT; U_{qt} =2.07kT; $U_{\tilde{\sigma}q}$ = -2.24kT; K_{σ} =1.46mmHg^{-1};
$K_{\tilde{\sigma}}$ =1.14 M^{-1}.

B.   Data points [9] taken with 1.17×10^{-4}M hemoglobin (by tetramer)
and BisTris buffer.   Total DPG: △ none; ▢ 2.87×10^{-5} M;
◇ 5.75×10^{-5} M; ○ 8.62×10^{-5} ; ⊠ 1.15×10^{-4} M; ▽ 1.15×10^{-3} M.

Theoretical equilibrium curves shown for parameter values:
U_q = -4.61kT; U_{qt} =1.19kT; $U_{\tilde{\sigma}q}$ = -3.94kT; K_{σ} =.236mmHg^{-1};
$K_{\tilde{\sigma}}$ =1.39 $\times 10^5$ M^{-1}.

Here K_{σ} and $K_{\tilde{\sigma}}$ are respectively proportional to exp $(-U_{\sigma}/kT)$ and
exp $(-U_{\tilde{\sigma}}/kT)$; kT=2.5 $\times 10^{-2}$ev at 20° C.   Induced-fit binding (U_t = ∞
and $U_{\sigma t}$ = ∞) and zero t-t interactions (U_{tt} = 0) were assumed.

shown in Fig. 12. The prediction that the change from the deoxy to
the oxy quaternary structure occurs at about the third oxygenation,
and that it is not very sensitive to the DPG concentration, is in agreement with experimental results [8:5521].

THE MWC MODEL AND HEMOGLOBIN OXYGENATION

Although the mechanism of cooperativity in hemoglobin is thought to
be more accurately described by the Perutz picture, the MWC model
is known to fit oxygenation curves very well. This is as it should be,
since the MWC and Perutz pictures are equivalent with respect to
oxygenation (i. e. substrate binding at different sites is interrelated
in the same way in the two cases) as may be seen by looking at the

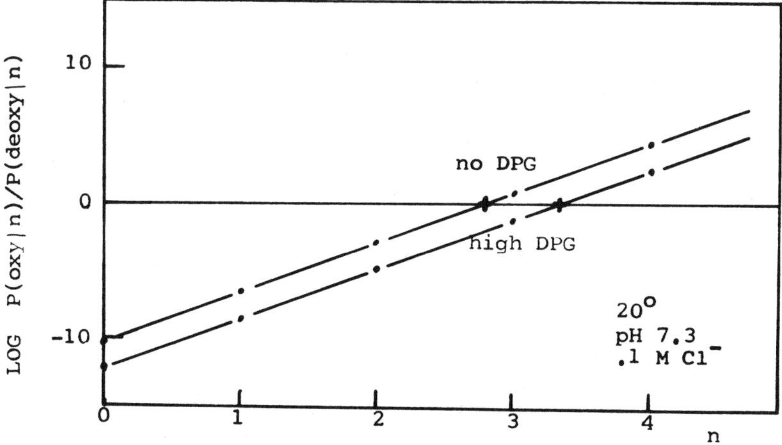

Figure 12.   Theoretical calculation of the relative probabilities of
the oxy and deoxy quaternary states in a hemoglobin molecule with n
hemes oxygenated (P(oxy|n) and P(deoxy|n) respectively). Theoreti-
cal parameters as in Fig. 1A. The "change" from one quaternary
conformation to another is marked by a '+'.

relationships between the boxes in the diagrams in Fig. 9. The only
difference between the two models is that for MWC the tertiary
structure is determined by the quaternary structure (due to the
symmetry assumption), and for Perutz the tertiary structure is
determined by substrate binding (due to the induced-fit assumption).

ACKNOWLEDGMENT

The first author is grateful to Professor Robert J. Silbey for his
generous moral and financial support of this work. We are much
obliged to Mr. Arnold Reinhold for his painstaking and astute guid-
ance in the programming aspects of this endeavor, and for the loan
of an excellent maximizing routine. Thanks are also due to Professor
H. Franklin Bunn [9] for taking a new set of data for our use, and to
Professors Ruth and Reinhold Benesch [10:2567] for providing their
raw data. This work was supported by ARPA Grant SD-90 under the
auspices of the MIT Center for Materials Science and Engineering
and by NSF Grant GP-15428.

REFERENCES

[1]   Torrance, J. D., C. Lenfant, and G. A. Finch. 1969. Fors-
varsmedicin . 5:187.

[2]   Perutz, M. F. 1970. Nature. (London). 228:726.

[3]   Herzfeld, J. 1972. Ph.D. Thesis, MIT.

[4]   Pauling, L. 1935. Proc. Natl. Acad. Sci. U.S. 21:186.

[5] Monod, J., J. Wyman, and J.-P. Changeux. 1965. _J. Mol. Biol._ 12:88.

[6] Koshland, D. E., G. Nemethy, and D. Filmer. 1966. _Biochemistry._ 5:365.

[7] Thompson, C. J. 1968. _Biopolymers._ 6:110.

[8] Gibson, Q. H. and L. J. Parkhurst. 1968. _J. Biol. Chem._ 243:5521.

[9] Bunn, H. F. Unpublished data.

[10] Benesch, R. E., R. Benesch, and C. I. Yu. 1969. _Biochemistry._ 8:2567; Benesch, R. E. and R. Benesch, private communication.

CHAPTER 5

ELECTRONIC STRUCTURE AND FUNCTION OF METAL IONS IN BIOMOLECULES

Keith H. Johnson

INTRODUCTION

Metal ions in relatively small concentrations are known to be vital to
the metabolic processes of the living cell, e. g. , as the active centers
of certain enzymes and proteins, and as charge carriers in transport
processes through cell membranes. Among these, the most familiar
perhaps are the iron-containing hemoproteins, including hemoglobin,
myoglobin, cytochrome, catalase, and peroxidase.

Hemoglobin, found in the erythrocytes and responsible for the transport of oxygen between the lungs and tissues, is the most thoroughly
investigated hemoprotein. The transport of oxygen occurs through the
reversible chemisorption of oxygen molecules by an Fe atom
strategically located in prosthetic porphyrin-like heme groups attached
to the protein macromolecule. A similar hemoprotein, myoglobin, is
responsible for storing oxygen in muscle tissue. The cytochromes
are part of the respiratory chain of enzymes which are responsible for
the oxidation of various substances by living cells. This oxidation is
not carried out by oxygen directly but by the catalytic transfer of
electrons from the substance to oxygen via the cytochromes. That
part of the cytochrome where the "redox" process is occurring is an

iron-porphyrin heme group. Catalase and peroxidase are enzymes which catalyze the oxidation of special substances by hydrogen peroxide, again through the electronic activity of Fe atoms located in porphyrin-like prosthetic groups attached to the protein macromolecule. Enzymes and proteins containing the metals cobalt (e. g., vitamin B_{12}), zinc (carboxypeptidase), copper (ceruloplasmin), and magnesium (chlorophyll) are also recognized to be biocatalysts of important metabolic processes occurring in the living cells of animals and plants.

The biological functions of these systems cover a wide spectrum, varying with the particular metal and organic ligands with which it interacts. Nevertheless, there are gross similarities in the stereochemical geometries of the active centers of these macromolecules, e. g., approximate octahedral, tetrahedral, or square-planar coordination of the organic ligands of the prophyrin-like prosthetic group with a central metal atom (see Fig. 1). A quantitative experimental and theoretical analysis of the electronic structures and chemical bonding of these metals in their local molecular environments should lead to a better understanding of the nature of their biocatalytic and transport functions.

The problem of metal-ion (e. g., sodium, potassium, magnesium, calcium) transport through membranes appears, at first sight, to be unrelated to that of metal ions in the active centers of enzymes. The latter problem is distinguished by a rather specific chemical bonding

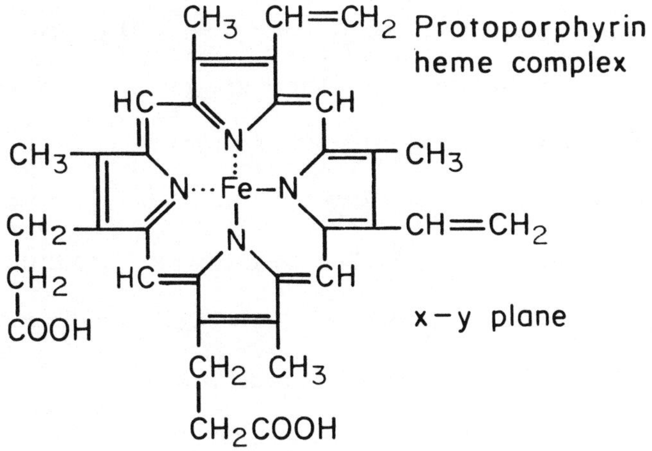
Protoporphyrin heme complex
x−y plane

(a)

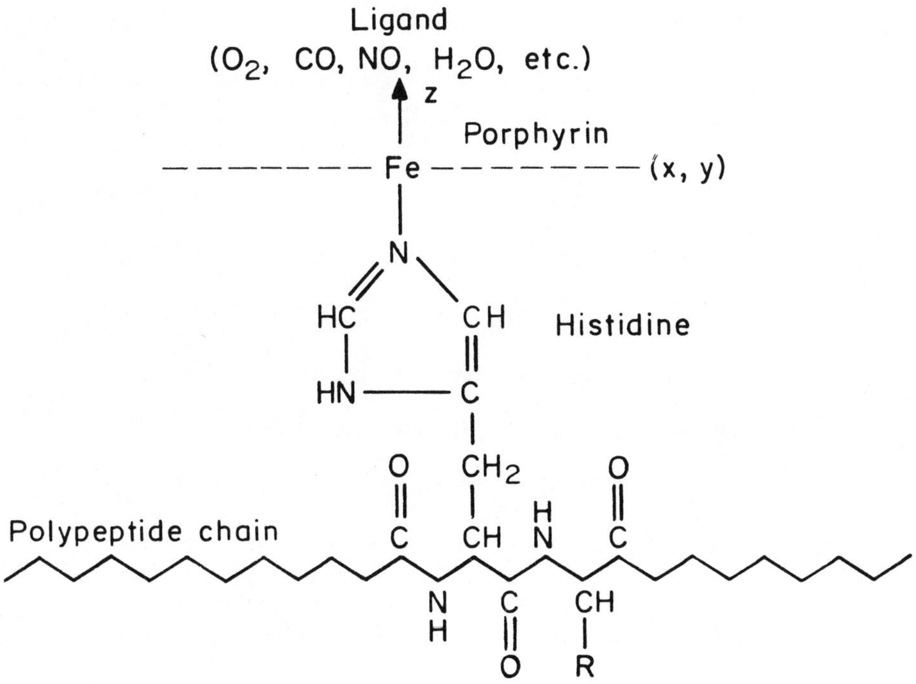
Ligand
(O_2, CO, NO, H_2O, etc.)
z
Porphyrin
(x, y)
Histidine
Polypeptide chain

(b)

Figure 1

of metal ion to ligands, while the former is characterized by a selec-

tive but high mobility of the metal ion through the membrane. Never-

theless, there is a close relationship of the biocatalytic function of

enzymes with membrane structure and transport, since many enzymes

are believed to be bound to the cellular membrane interfaces. The act

of biocatalysis may be facilitated by changes in the basic membrane

network geometry and in the cation distributions.

Both the problems of metallo-enzyme catalysis and membrane

specificity can benefit by one's experience with somewhat analogous

non-biological problems, e.g., the catalysis by transition metals in

industrial applications and the specificity of ordered and amorphous

crystalline lattices for certain metal-ion "impurities." Research on

the electronic structures of metal ions in biomolecules and in bio-

materials is essential for the development of an understanding of both

their natural biological functions in living cells and their interaction

with prosthetic materials. Such research can lead to a better under-

standing of the nature of normal and abnormal physiological enzyme

activity, and will be relevant to the question of how metal ions and

enzymes interact with natural and synthetic membrane materials.

THE HEMOPROTEINS

Of all the naturally occurring enzymes and proteins containing metals,

the hemoproteins have received the greatest amount of attention. As

is well known, protein molecules are long-chain polymers consisting

of amino-acid residues called polypeptides, each protein being characterized by a specific sequence of amino acids along the polypeptide
chains. The spatial configurations of polypeptide chains in hemoglobin
and myoglobin in crystalline form have been investigated thoroughly
via X-ray methods by Kendrew, et. al. [1:422] and Perutz [2:646] .
The hemoglobin molecule, for example, consists of 574 amino acids
($\sim 10,000$ atoms) organized into four intricately wound chains. A
prosthetic heme complex is tucked into each of the four chains. The
heme group itself consists of an iron ion coordinated by four nitrogen-
atom ligands which are part of a planar molecule known as protoporphyrin (See Fig. 1(a)). A fifth nitrogen ligand "below" the plane of
the porphyrin attaches the heme complex to the protein chain via the
amino acid histidine (see Fig. 1(b)). The corresponding position
"above" the plane of the porphyrin is of paramount importance to the
biological function of hemoglobin, for it is here that an oxygen molecule (or other ligands such as carbon monoxide) can be attached to the
iron ion for transport between the lungs and tissues. If an oxygen
molecule attaches itself by only one of the oxygen atoms, then this
atom together with the five nitrogen ligands are in six-fold coordina-
ation with respect to the central Fe ion [3] (see Fig. 2). It has also
been suggested that the O_2 molecule may be joined symmetrically to
the heme by both O atoms, leading to a seven-fold geometry [4:23] .
From steric considerations, some conformational changes of the heme
complex during oxygenation can be expected, including a small dis-

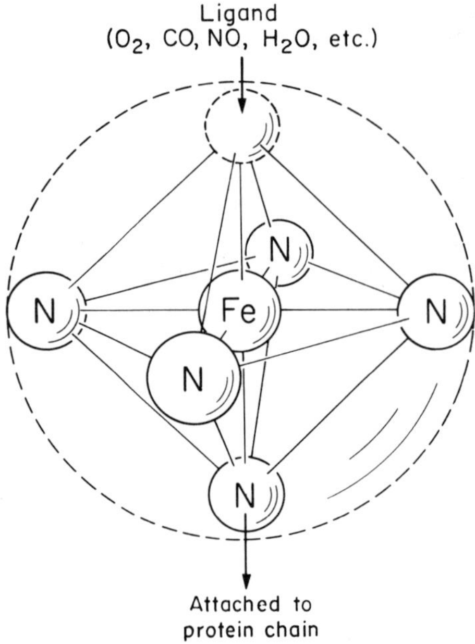

Figure 2

placement of the iron atom into the plane of the porphyrin [5] . The

other hemoproteins - myoglobin, cytochromes, catalase, and peroxi-

dase - have prosthetic porphyrin heme complexes similar to that

shown in Fig. 1.

Thus, the biological activities of hemoproteins and probably most

other metallo-enzymes depend rather critically on the electronic

structure of the central metal ion as it is chemically coordinated with

the surrounding ligands. When the ions are transition metals, such as

iron or cobalt, the magnetic "spin polarization" of the unfilled d-

electron shells is of primary significance. Most of our present under-

standing of the electronic structure of the hemoproteins at the experi-

mental level is based on magnetic susceptibility data [6:232] , electron paramagnetic resonance [7:67] , Mössbauer measurements [8:3], and

nuclear magnetic resonance [9:43] . Theoretical calculations have

been limited almost exclusively to the use of ligand-field methods

[10:159] which unfortunately provide no information about electronic

"charge-transfer" effects in the metal-ligand chemical bonds.

Under normal physiological conditions the Fe ion in hemoglobin and

myoglobin is in the divalent, ferrous Fe^{2+} electronic state. In the

ferric Fe^{3+} state, the ability of the Fe ion to transport or store oxygen

is destroyed. The change of the color of blood from dark red to bright

red is an indication of the effects of oxygenation on the electronic

structure of the heme. On the basis of magnetic measurements such

as those mentioned above, it has been established that the heme group

is paramagnetic in deoxyhemoglobin and diamagnetic in oxyhemoglobin.

There are six valence electrons in the free Fe^{2+} ion, all in the five

available orbitals of the 3d shell. Each orbital is capable of holding

two electrons, consistent with the Pauli exclusion principle. However,

the electrons tend to repel each other, so before any orbital can be

doubly occupied it is energetically favorable for each to be singly

occupied first. Consistent with this picture and with the measured

paramagnetism, the spin state of the Fe^{2+} ion in deoxyhemeglobin is

$S = 2$. The diamagnetism of oxyhemoglobin suggests spin pairing upon

oxygenation, i. e. , $S = 0$. When the O_2 molecule is chemically bound to

the iron ion, the repulsion of the electrons within each orbital is

exceeded by the repulsion between the iron and oxygen electrons, thus

allowing pairing in the same orbital. Similar simple arguments may
be used to describe the electronic states of iron in the other hemoproteins. For example, during the catalytic activity of the cytochromes, the iron ion seems to be changing between the ferrous and
ferric states.

Although considerable research has been carried out on the hemes,
a quantitative relationship between their electronic structures and
biological functions has not yet been established. For example, one
of the most interesting physiological properties of the hemoglobin
molecule is the so-called "heme-heme interaction." This interaction
reflects the fact that the binding of oxygen at any one heme group
increases the affinity for oxygen at the remaining groups. The iron
ions of the hemoglobin molecule are too much far apart (25-35A) to
allow for direct interaction. This is not an isolated phenomenon. In
a large number of enzymes there are several active sites, i. e. , locations in the molecule which interact with the substrate. When one site
is occupied by a substrate, the catalytic activity of the other sites is
influenced. The nature of this cooperative phenomenon among prosthetic groups and its relationship to biological function is not well
understood. It is likely that the cooperative effect is due, at least in
part, to conformational changes produced in the active sites during
biochemisorption and biocatalysis, and then "propagated" along the
protein chains to neighboring active sites. A detailed analysis of the
electronic structures of these active sites is essential for a quantita-

tive description of such effects.

METAL-ION TRANSPORT THROUGH MEMBRANES

The mechanisms of transport through cell membranes are perhaps
even less well understood than the activities of metallo-enzymes.
Much must still be learned about the factors governing so-called
simple passive diffusion through living membranes, not to mention
the more difficult problem of active transport. The physical forces
which are significant to membrane transport, namely (1) concentration
differences, (2) differences in activity coefficients, (3) electric potential differences, and (4) solvent "drag" forces, may operate individually or in complex combinations. There have, thus far, been essentially no fundamental investigations into the relationship, if any, between
the electronic structures of metal ions in the local environment of the
membrane and transport properties such as membrane specificity.

For many years there have been several competing models for the
structures of cell membranes, including (1) the lipid pore models
[11:385] , (2) homogeneous models [12] , and (3) bilayer models
[13:979] . Theoretical work on the mechanisms of passive transport
has been restricted largely to the theory of rate processes and
irreversible thermodynamics [14] . With respect to active transport,
a variety of mechanisms have been proposed, including (1) the fluid
circuit mechanism [15:141] which is basically a selective pore transport model, (2) simple membrane carrier mechanism [16] in which

the substance transported is presumed to form a complex with a membrane constituent, (3) the electron-linked mechanism [17] where it is assumed that an essential step in the transport is an electron transfer, and (4) the "propelled carrier" mechanism [18:1] in which the complex is drawn through the membrane by the contraction of a contractile fiber. Eisenman [19] has developed a theoretical model, based largely on thermochemical arguments, for the biological specificity of cell membranes for particular ions and the effects of these ions on membrane permeability.

It is likely that certain concepts of solid-state physics and materials science will be useful in developing a better understanding of biological membrane transport. Consider the well known problem of membrane specificity for sodium and potassium ions. In a somewhat analogous nonbiological problem, considerable differences are found in the nature of the coordination properties of sodium and potassium, respectively, for oxygen in crystalline solids. Sodium can form only structures with coordination numbers of six and eight with respect to oxygen. However, coordination numbers of potassium for oxygen of six, seven, eight, nine, ten, and twelve have been found. Thus, although sodium requires a rather restricted spatial arrangement of oxygen ligands, there is a relatively wide range of oxygen structures which can accommodate potassium ions. It is possible that similar arguments, based on the nature of local structure and coordination in amorphous biological membranes, may be applied usefully to the problems of specificity in sodium and potassium transport.

One of the most exciting recent developments in the field of metal-ion transport through membranes is the discovery of a large class of macromolecules which are capable of making both natural and synthetic phospholipid membranes selectively permeable to metal ions, particularly potassium (K^+) [20], [21:173]. These substances include macrocyclic peptides and depsipeptides such as the antibiotic valinomycin, as well as certain "pseudocyclic" linear peptides. There is structural evidence (based on X-ray diffraction data) that these macromolecules contain the metal ions in "molecular cages" (with ligand coordinations somewhat like those characterizing the hemes) and then act as "carriers" of the ions through the membrane. Although it is suspected that the ion is held within the cage via electrostatic "ion-dipole" interactions, the precise nature of the bonding is unknown. Evidently local stereochemical conformation is very critical to the observed specificity of these molecules for particular ions.

THEORETICAL CALCULATIONS OF ELECTRONIC STRUCTURE

Most of the difficulties in developing a fundamental understanding of the biological activities of metal-ion complexes in terms of their chemical bonding are due to basic defects associated with the commonly adopted theoretical techniques. Ligand-field (or crystal-field) methods [10] are based on the assumption that the central metal-ion is held together by electrostatic force, and that the behavior of the ion depends solely on the electric field from the surrounding ligands. While ligand-field theory has led to some useful information about the electronic configurations of iron in the hemes, "charge-transfer" and other important electronic delocalization effects involving the metal-ion and surrounding ligands are completely neglected. To consider such effects, a full molecular-orbital theory is necessary. Unfortunately, conventional Hartree-Fock molecular-orbital techniques [22] based on representing the electron wavefunctions as linear combinations of atomic orbitals (LCAO methods) are difficult and costly in computer time to implement on complex, organo-metallic molecules because of the necessity of having to compute millions of multicenter integrals or equivalent Hartree-Fock matrix elements. Simpler semi-empirical LCAO-type molecular-orbital methods, such as those based on the "complete neglect of differential overlap" (CNDO methods) [23] depend on the ad hoc (often not physically justified) parametization of matrix elements and yield only semiquantitative results for transition-metal systems.

In recent publications [24:3085] , [25:844] , [26] we have developed and applied a new theoretical technique, in the form of a spin-unre-stricted self-consistent-field cluster model, for describing the chemi-cal bonding of metal ions and other atoms in complex polyatomic molecules and crystals. This model permits the accurate calculation, from first principles, of electronic energy levels and wave functions for the cluster, but requires relatively little computer time. The cluster, which may be a free polyatomic molecule, part of a biological

macromolecule, or a polyatomic complex in crystalline or membrane-

like matrix, is geometrically partitioned into contiguous <u>atomic</u>, <u>inter-</u>

<u>atomic</u>, and <u>extramolecular</u> regions. The one-electron Schrödinger

equation is numerically integrated within each region in the partial-

wave representation for spherically averaged and volume averaged

potentials which include Slater's [27] Xα statistical approximation to

exchange correlation. The wave functions and their first derivatives

are joined continuously throughout the cluster via multiple-scattered-

wave theory. The effects of a particular biomolecular environment on

the cluster are described by boundary conditions, e. g. the matching

of the solutions of Schrödinger's equation in the extramolecular region

to those within the atomic regions at an artificial spherical boundary

surrounding the entire cluster. This numerical procedure is repeated,

using the wavefunctions obtained at each iteration to generate a charge

density and new potential, until self consistency is attained.

In our applications of this technique to the electronic structure of

metal ions in the prosthetic groups of enzymes and proteins (e. g., the

heme clusters) we are attempting to determine:

(a) orbital and spin configurations of the metal-ion electrons,

especially the d-electrons

(b) the importance of "charge transfer" between metal-ion

and surrounding ligands

(c) the existence of low lying excited molecular orbitals

(d) contour maps of the electronic charge and spin densities

   throughout the heme clusters.

On the basis of these calculations, we hope to arrive at "energetic"

and "structural indices" which are related to the physiochemical

properties and biological activities of the molecules, e.g., the most

energetically favorable geometry of the heme complex. Finally, we

hope to develop a quantitative theory for the relationship of the elec-

tronic structures of metal ions in metallo-enzymes to their functions

in biochemisorption and biocatalysis.

Along with this research on metal ions in enzymes and proteins, we

are also applying the cluster theory to the electronic structure of

metal ions undergoing transport through natural and synthetic mem-

branes. Particular emphasis is being placed on determining the

nature of the bonding of metal ions in those "carrier" macromolecules

discussed earlier which are responsible for enhancing the selective

permeability of phospholipid membranes. We hope to establish the

relative importance of "ion-dipole" and "chemical bonding" inter-

actions between the metal ion and those atoms which encage the ion in

the carrier molecule. We are also investigating the relationship

between local ion coordination number and membrane specificity.

REFERENCES

[1] Kendrew, J.C., R.E. Dickerson, B.E. Strandberg, R.G. Hart, D.R. Davies, D.C. Phillips and V.C. Shore. 1960. Nature. 185:422.

[2]   Perutz, M.F. 1965.   J. Mol. Biol. 13:646.

[3]   Pauling, L. 1949.   The Electronic Structure of Haemoglobin,
In Haemoglobin, p. 57. London: Butterworths Scientific Publ. ; 1964.
Nature. 203:182.

[4]   Griffith, J.S. 1956. Proc. Roy. Soc. (London), A235:23.

[5]   Hoard, J.L. 1966.   Sterochemistry of Porphyrins, In Hemes
and Hemoproteins, eds. B. Chance, R.W. Estabrook, and T.
Yonetani, p. 9.   New York: Academic Press.

[6]   Scheler, W. , G. Schoffa, and F. Jung. 1957.   Biochem. Z.
329:232.

[7]   Bennett, J.E. , J.F. Gibson, and D.J.E. Ingram. 1957. Proc.
Roy. Soc. (London). A240:67.

[8]   Lang, G. and W. Marshall. 1966.   Proc. Phys. Soc. (London).
87:3.

[9]   Wüthrich, K. and R.G. Shulman. 1970.   Physics Today. No. 4.
23:43.

[10]   Kotani, M. 1964.   In Advances in Chemical Physics, ed. I.
Prigogine, Vol. VII, p. 159.   London: Interscience Publishers.

[11]   Collander, R. 1937. Trans. Faraday Soc. 33:385.

[12]   Davson, H. and J.F. Danielli. 1943.   The Permeability of
Natural Membranes. Cambridge: Cambridge University Press.

[13]   Mueller, P. , D.O. Rudin, H. Ti Tien, and W.C. Wescott,
1962. Nature. 194:979; 1963. J. Phys. Chem. 67:534.

[14]   Stein, W.D. 1967.   The Movement of Molecules Across Cell
Membranes. New York: Academic Press.

[15]   Ingraham, R.C. , H.C. Peters, and M.B. Visscher. 1938.
J. Phys. Chem. 42:141.

[16]   Osterhout, W.J. 1906. Bot. Gas. 42:127; 1907. 44:259; 1914.
58:178.

[17]   Conway, E.J. 1953.   The Biochemistry of Gastric Acid Secretion. Springfield, Illinois : Thomas.

[18]   Danielli, J. F. 1954. Symp. Soc. Exp. Biol. 6:1.

[19] Eisenman, G. 1961. In Symposium on Membrane Transport and Metabolism, eds. A. Kleinzeller and A. Kotyk. New York: Academic Press.

[20]   Pressman, B. and D. H. Haynes. 1969. The Molecular Basis of Membrane Function, ed. D. C. Tosteson, p. 221. Englewood Cliffs, New Jersey: Prentice-Hall, Inc.

[21] Shemyakin, M. M. , V. K. Antonov, L. D. Bergelson, V. T. Ivanov, G. G. Malenkov, Y. A. Ovchinnikov, and A. M. Shkrob. 1969. The Molecular Basis of Membrane Function, ed. D. C. Tosteson, p. 173. Englewood Cliffs, New Jersey: Prentice-Hall, Inc.

[22]   Slater, J. C. 1963. Quantum Theory of Molecules and Solids. Vol. 1. New York: McGraw-Hill Book Company.

[23]   Pople, J. A. and D. L. Beveridge. 1970. Approximate Molecular Orbital Theory. New York: McGraw-Hill Book Company.

[24]   Johnson, K. H. 1966. J. Chem. Phys. 45:3085.

[25]   Slater, J. C. and K. H. Johnson. 1972. Phys. Rev. B5:844.

[26]   Johnson, K. H. Advances in Quantum Chemistry. Vol. 7, ed. P. -O. Löwdin. New York: Academic Press. (In press.)

[27]   Slater, J. C. Advances in Quantum Chemistry. Vol. 6, ed. P. -O. Löwdin, p. 1. New York: Academic Press.

PART II

PROBES THAT LINK THE MICROSCOPIC WITH THE MACROSCOPIC

CHAPTER 6

MOLECULAR MICROSCOPY

John G. King and James C. Weaver

INTRODUCTION

If an object is placed in vacuum, each square centimeter of its surface
both receives and emits neutral atoms and molecules, depending on the
surrounding pressure and the vapor pressure of the surface, according
to the equation $\dot{n} = 3.5 \times 10^{22} Kp/\sqrt{MT}$, where K is the evaporation
coefficient, p is the pressure in Torr, M is the molecular weight, T
the absolute temperature, and n is then given in atoms or molecules
per second. With water molecules at 300°K and 10^{-4} Torr, the
flux at the surface is 5×10^{16} molecules/sec cm^2 or 5 molecules/sec $\mathring{A}^2$.
The rate at which neutral atoms are emitted depends on temperature
and vapor pressure in an equilibrium situation However, molecules
can be caused to evaporate by a variety of stimuli, some of which
amount to heating the surface locally, while others may act essentially
directly on the bonds binding the molecules to the surface, as is the
case with such mechanical stimuli as rubbing, bending, or vibration
(ultrasound).

What binds molecules to surfaces? Roughly speaking, if it were not
for rotation most molecules would exhibit electric dipole moments of
the order of 1 Debye $(10^{-18} esu)$; so if a molecule is near the surface

of dielectric constant ϵ it can be imagined to see an image molecule in that surface ($\epsilon-1$) times in strength. The interaction will be of the form $p^2(\epsilon-1)r^{-3}$ so that for a reasonable distance between the surface and the molecule of the order of 1 Å, we see that the energies are of the order of 1 electron volt, 10^4 degrees Kelvin, or 25 kilogram calories/mole. This crude calculation serves only to remind the reader that interactions between neutral molecules are comparable in energy to thermal energy and give rise to the weak forces that are so important in chemistry and biology.

It seems possible that complex patterns of weak forces at surfaces are partly responsible for the highly specific sites at which molecules can bind, but at present there are no direct ways of studying such phenomena. Instead, they must usually be inferred by combining biochemical methods with X-ray diffraction analysis. However, by using the technique for handling neutral atoms and molecules developed over the years in connection with molecular beam spectroscopy and similar studies, we hope to develop a variety of instruments that will make it possible to study in a very direct way the interactions between neutral molecules and surfaces with spatial resolution approaching in some cases that obtainable with the electron microscope [1]. For convenience, we call all devices in which neutral molecules carry information from the surface to be studied, molecular microscopes, independently of the nature of the stimulus that causes the neutral atoms or molecules to leave the surface.

MOLECULAR MICROSCOPES AND EXPERIMENTS

"Self-Luminous" Sample

We can distinguish a number of kinds of molecular microscopes

depending on the arrangement of apparatus. Thus, for instance, it is

possible to build a device (Figure 1) with no spatial resolution in order

to analyze the transmission or emission of gases through or from

biological or medical samples, thus learning about diffusion on a

small scale of gases that are not ordinarily accessible to chemical

analysis in the usual arrangement. At the same time, much faster

time response is attainable, e. g. , on the order of milliseconds or

less as opposed to seconds or more in conventional methods. Thus

some information concerning the dynamics of the diffusion may be

obtained.

A somewhat more sophisticated apparatus provides some spatial

resolution (currently $\sim 100\mu$), and is expected to uncover new informa-

tion concerning regions that emit more or less gas and therefore have

functions or structure not hitherto appreciated [2] . Such experiments

are underway in our laboratory at this time and consist, for example,

of mounting a frog skin in contact with appropriate aerated Ringer' s

solution and then observing the neutral emissions in the vacuum region

above it by converting the molecules into ions, eventually using a mass

spectrometer to determine what species are involved. A view of such

an apparatus is given in Figure 2 with a fairly detailed caption explain-

ing the functions of the various components. Some of the samples

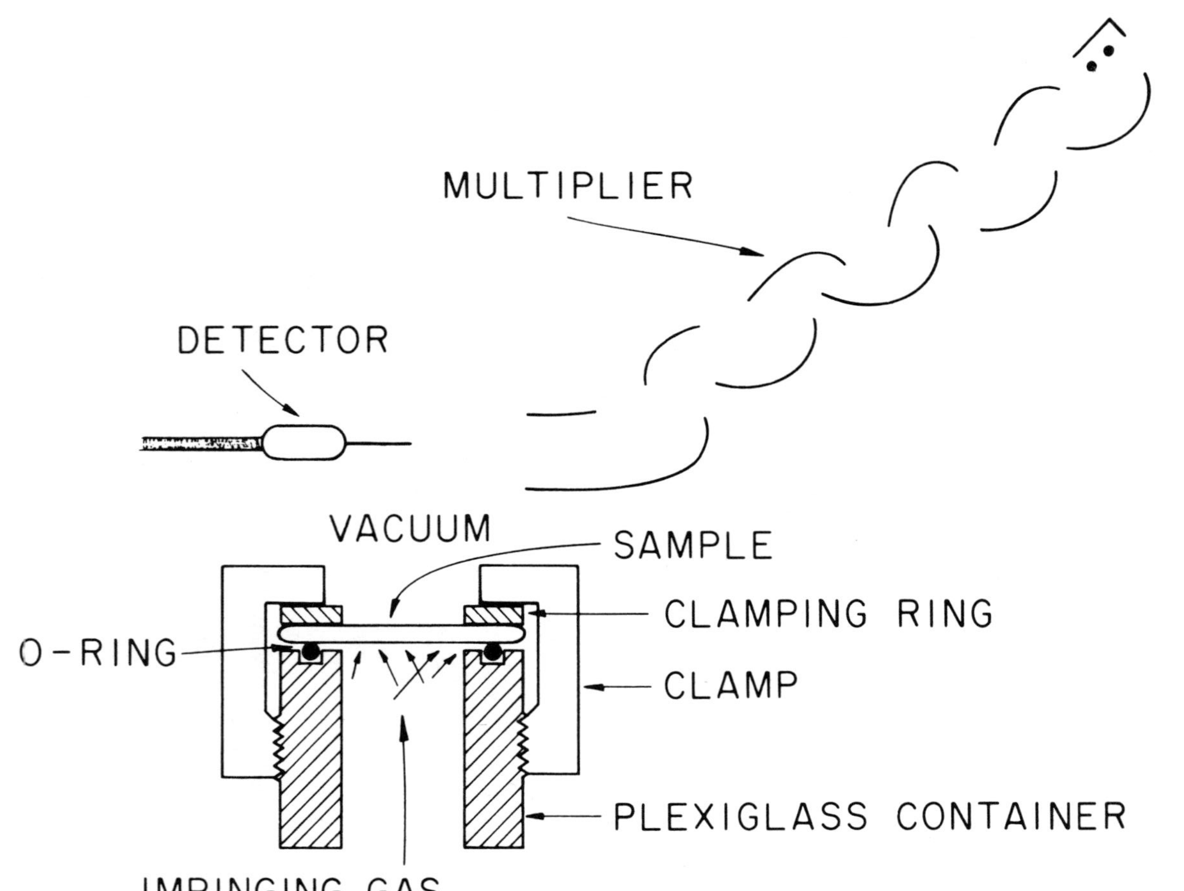

Figure 1.   Schematic of an apparatus for studying transmission of gases through thin samples or membranes without spatial resolution.

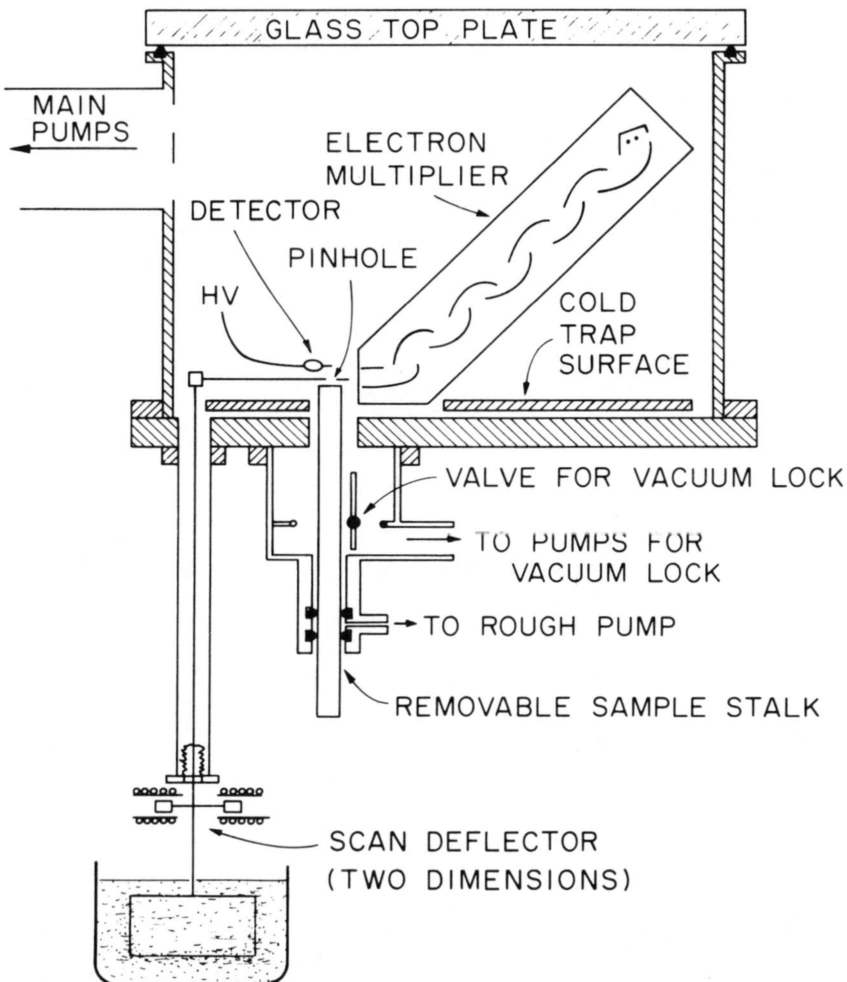

Figure 2.   Schematic of the scanning water molecular microscope, which is a "pinhole" microscope. This device has an initial resolution of about 100μ and produced the picture shown in Figure 4. A scan signal drives both the scan deflector (shown) and an oscilloscope. The brightness of the oscilloscope trace is modulated by the electron multiplier signal so that a visual picture of the variation in the flux of neutral water molecules from the sample is obtained. This apparatus has a vacuum lock so that removable sample stalks can be introduced, and has a pumpdown time of about one hour. A variety of sample holders and samples can be mounted on the top end of the sample stalk. Typical scan times are 3 to 6 minutes.

under consideration for study by this method follows: membranes, microtome samples of various organs and tissues, nerves, spores, vegetable material of various kinds. It is also possible to observe the effect of various stimuli, in particular, electric currents and hormones in the case of membranes, light in the case of photosensitive materials and lastly, electric impulses in the case of nerves [3:921].

Excessive denaturing of the sample by drying in the vacuum on one side can be combated by interposing millipore filter or other material, sacrificing some spatial and temporal resolution in order to maintain a higher level of moisture in the outer layers of the sample. It is also possible to maintain a general water vapor atmosphere provided it does not interfere with analyzing the emissions that are to be studied. In this way the sample can be kept from dessicating entirely, in that a monolayer or so of water would remain on the surface. For instance, if it were desired to study metabolism by monitoring carbon dioxide emission, the presence of water vapor would maintain the moistness of the sample. Provided that the density of water vapor was not so high as to appreciably shorten the mean free path, neither the spatial or temporal resolution should suffer, and a properly designed mass spectrometer could readily discriminate against the large amount of water ions derived from the background vapor.

A recent picture obtained with the apparatus of Figure 2 is shown in Figure 4. Here, as depicted in Figure 3, the specimen consists of (A) a plexiglass cover containing a 1/16" hole exposed to the vacuum region of the molecular microscope. (B) is an interface consisting of

nucleopore filter material (1. 0 μ average pore size) and (C) is (dead)

frog skin which is oriented such that the natural exterior faces the

vacuum. The inside of the skin is bathed with water.

Since the frog skin is permeable to water over the entire area and

not just the area exposed by the 1/16" hole, one might expect water to

cross the frog skin and migrate laterally in those regions covered with

plexiglass. This lateral migration should feed the circumference of

the 1/16" hole, with increased emission from the circumference

region. This possibility is in fact confirmed by our observations,

which provide unprecedented direct pictures of the water flux by using

the water molecules themselves to make a picture. A portion of the

circumference and the region inside the hole are shown in the molecu-

lar microscope picture in Figure 4. The magnification is about 20x.

Note the bright region <u>inside</u> the circumference of the hole. This

reproducible feature is apparently real, and presumably indicates a

region of greater water permeability of the (dead) frog skin. It is not

yet known what causes this; possibly this is an area of damaged tissue.

Stimulated Desorption

Another kind of instrument makes use of finely focussed charged

particle beams (electrons or helium ions) to scan and heat selectively

various parts of the surface, or to break weak bonds between mole-

cules and the surface. In either case, neutral molecules are desorbed

from the surface. For example, it is a common experience in electron

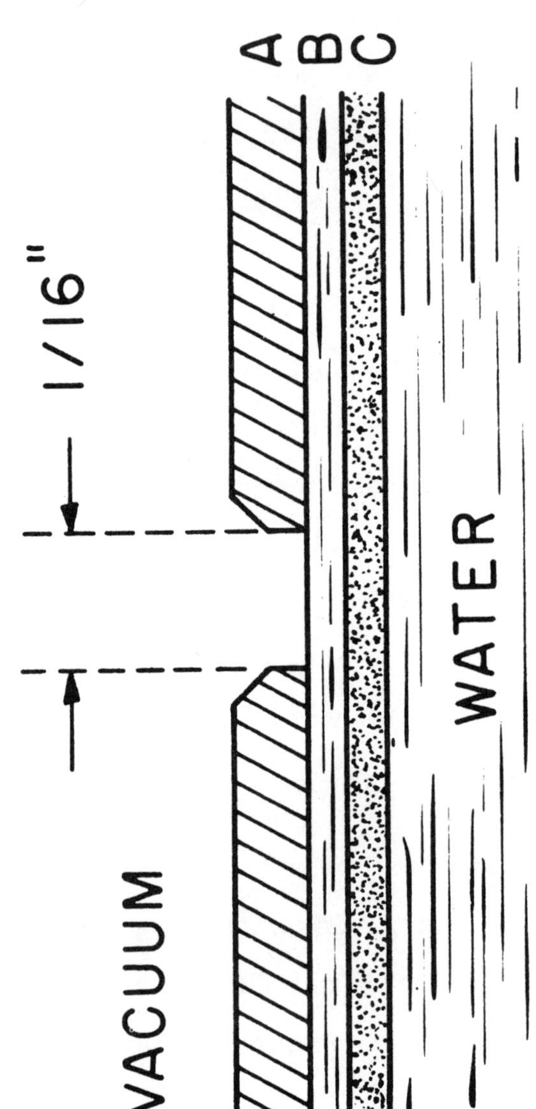

Figure 3.   Schematic detail of the sample holder and sample used for the picture shown in Figure 4; the natural exterior of the frog skin faces the vacuum.

Figure 4.   Picture of water transmission across a (dead) frog skin at
a magnification of about 20x.   Brightness corresponds to a larger
water flux, darkness to a lower one, with overall intensity levels
adjustable and somewhat arbitrary.   An interpretation of the structure
shown appears in the text.

microscopy to have outgassing of the sample and it is our plan to

observe and analyze these neutral emissions that are ordinarily so

troublesome.   Because there is no image charge, we anticipate that

such a system of microanalysis will be reasonably efficient in terms

of the number of neutrals coming off the surface per stimulating

electron or ion striking it.   Furthermore, the efficiency of detection

of neutrals, though far short of that for ions, is much higher than that

for X-rays so that our instrument, such as that shown in Figure 5 is

competitive with the various microanalytical devices that have been

developed.

By using a pulsed beam and waiting for the slow neutrals to get off

the surface and be detected, one can scan the entire surface with the

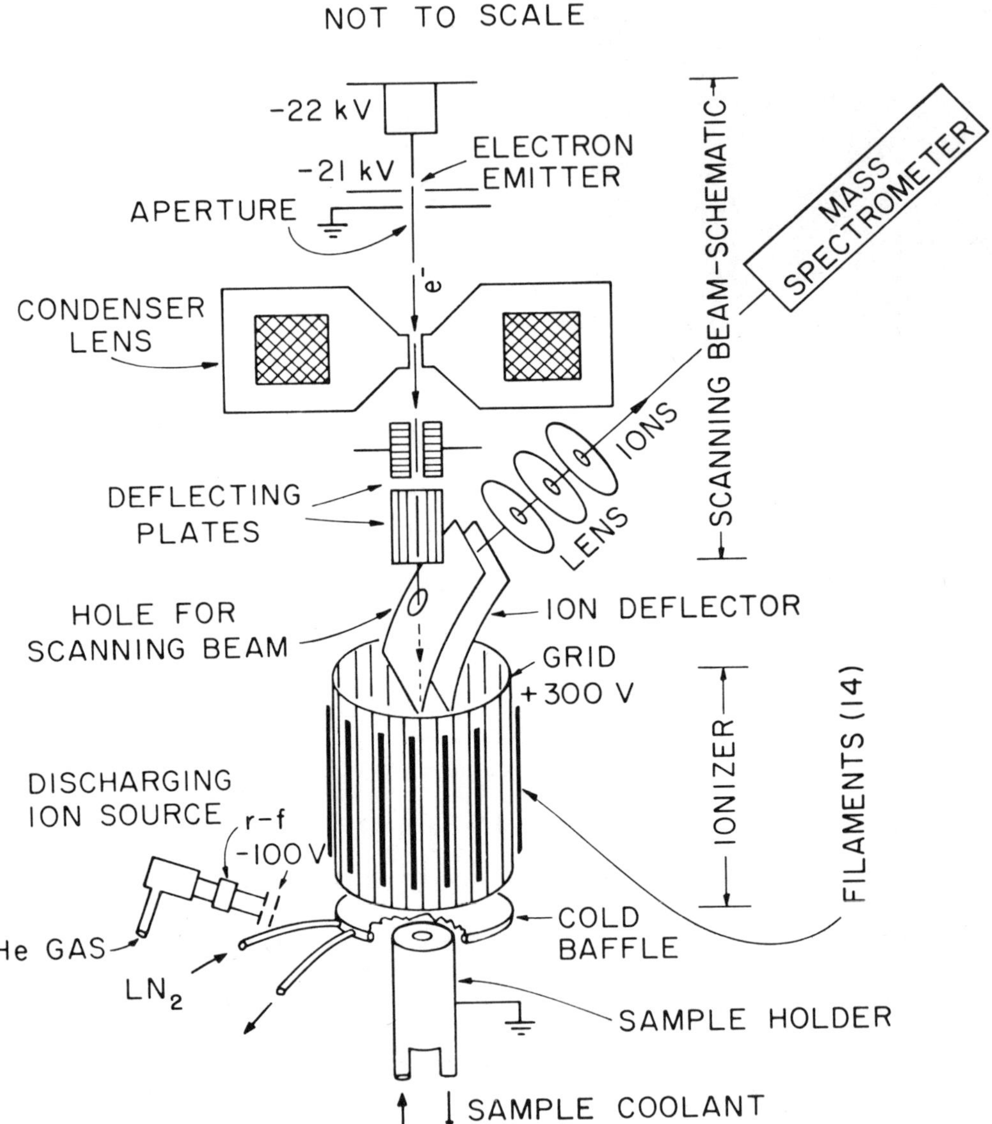

Figure 5.   Schematic (not to scale) of the stimulated desorption molecular microscope which is intended for thrombin and other studies. This apparatus has a potential resolution of a few angstroms. Its purpose and general method of operation are described in the text.

same resolution that is attainable in charged particle scanning devices (approximately 5 angstroms). The very high current density required by the scanning electron microscope is not likely to be necessary in this instance. An additional benefit exists in that the scanning electron beam makes it possible to eventually do simultaneous electron and molecular microscopy so that the features might be examined in a relatively conventional way and by the neutral emissions. Needless to say such an instrument would be fairly elaborate.

We have started the design of such an instrument that will be developed in the next few years for the purpose of finding thrombin molecules (molecular weight 36,000) adsorbed to surfaces such as polyethylene in an effort to understand the nature of the adsorption. Such adsorption may be of considerable medical importance in making non-thrombogenic surfaces [4]. We anticipate that there will be many microanalytical applications of an instrument of this sort. It is also possible that by irradiating the surface with neutral molecules of various kinds and then observing from where they can be unstuck by the scanning beam that one could investigate the nature of the bonding at the surface. Thus, one can imagine that water may stick at hydrophylic spots and carbon tetrachloride at hydrophobic points, but of course whether this would be true on a microscopic scale and with the sample more or less denatured as it would be in vacuum remains to be seen. Nonetheless, we believe that this is the instrument which can yield the most fundamental results in the shortest time.

Scanning Molecular Beam

Finally, it is possible to imagine an instrument in which a highly collimated beam of molecules scan the surface and in which one observes the scattered beam. Whether or not it is possible to make a beam of molecules with five to ten angstrom width is not yet clear. At least a few preliminary experiments are needed.

Lenses

Plainly the problem of lenses becomes increasingly difficult as one goes from light to charged particles to neutral molecules. In the last case the lenses are, of course, arrangements of inhomogeneous magnetic or electric fields, but it seems difficult to produce a lens for neutral molecules of short focal length and comparatively large aperture with existing techniques. This means that one must use a pinhole as the imaging arrangement with a consequent loss of solid angle. Pinholes can easily be manufactured down to fractions of a micron and apparently even less by ion track etching methods. But, of course, the difficulties because of fragility and with plugging plainly become very severe. On the other hand it is possible to investigate the performance of an electrostatic lens under a variety of assumptions with the result that an approximate formula for the focal length in terms of the diameter, the maximum voltage, and the polarizability, mass and velocity of the molecule can be derived.

$$f = d^3 V^{-2} \alpha^{-1} mv^2 \quad \text{(cgs)} \tag{1}$$

Unfortunately, these lenses will be both velocity sensitive and state sensitive. Nonetheless, this gives some indication of what may be achieved and of course a very bad lens would still be superior to a pinhole as far as atom gathering power was concerned. A quite radical suggestion has been advanced by Dr. T. R. Brown, in which a drop of superfluid helium at a suitably low temperature maintained in a small aperture through surface tension might exhibit focusing properties on account of the fact that m^* the effective mass of an impurity moving through superfluid helium is larger than the mass of the atom itself. Predicting the performance of such a lens requires more detailed understanding of what happens to energy and momentum at the surface of the drop, and, in fact, more information about m^* for foreign objects other than helium 3 and ions with vortices. Also, for most light molecules $v_{thermal} \simeq 5 \times 10^4 \, cm\text{-}sec^{-1} \gg v_{s,c} \simeq 7 \times 10^3 \, cm\text{-}sec^{-1}$, the intrinsic superfluid critical velocity, and it is not clear how significantly the trajectory of foreign molecules will be affected.

Molecular Photography

One partial solution to the pinhole problem (lack of intensity) would be to abandon scanning with its low duty cycle and take essentially snapshots by freezing the molecules on a clean surface and then reading them off at one's convenience at a later time. In this way, we estimate we could obtain 500 angstrom resolution with the molecular microscope now being used to study frog skin . Besides collecting the molecules on a cold surface for later desorption, which has many

obvious difficulties, it might be possible to develop a scheme of
molecular photography in which the molecules act as nucleating
centers for some distinctive chemical reaction, thereby taking advantage of the same immense chemical amplification that makes optical
photography so successful.

Flow Chart Comparison

The various twists and applications that molecular microscopy might
take are depected in the chart of Figure 6. Of these, several have
been discussed in the preceeding sections, and the few that have not
are self-explanatory.

SUMMARY

We can summarize the intentions of molecular microscopy in a variety
of ways. First of all, the interaction of the particles that carry the
information to the observer is quite different in the case of neutral
molecules than in the case of either photons or electrons. Instead of
being a direct electromagnetic interaction through induced dipole or
scattering by the shielded nuclear charge it involves the weak forces
at surfaces. Furthermore, it is very highly surface specific, since
diffusion effects of large neutral molecules from within the sample
take place slowly. The instrument also has a certain utility as a
microanalyzer. But none of these properties are as important as the
fact that it exploits the very same forces that are important in the

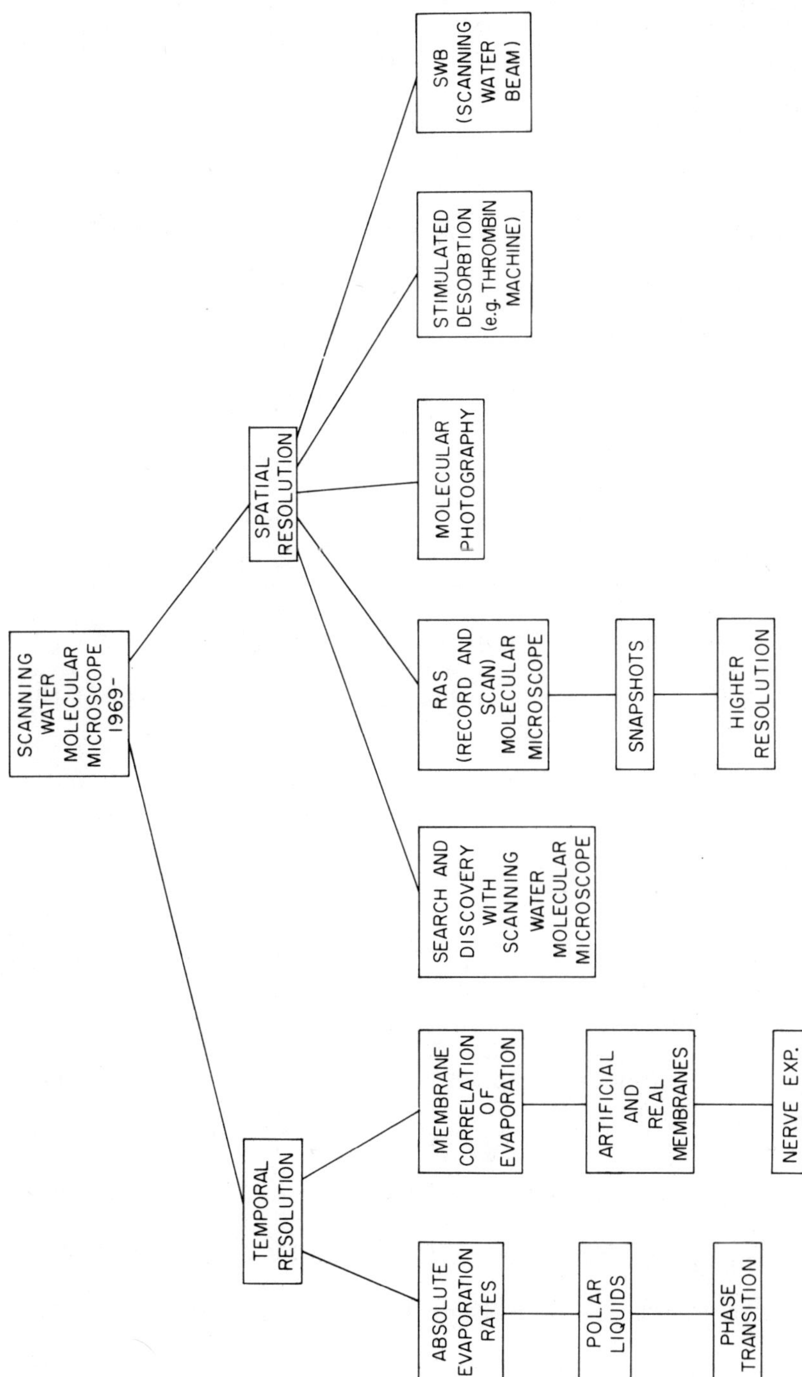

Figure 6.   Chart of various possiblities, applications and directions of the development of molecular microscopy.

dynamical situations of the sample. In this sense, even though the electron microscope continually improves it is nonetheless not going to provide the same kind of information that the molecular microscope does. Both the optical and electron microscope are basically low contrast devices and require staining in order to make various features visible, and the history of such microscopy is inexplicably intertwined with the history of staining. Even when that Holy Grail of the electron microscopists is attained, the direct visualization of biologically significant molecules, it will still not tell us directly about the weak forces that govern the bio-chemistry of the sample. We feel that this is where the molecular microscope will make an important contribution. There is certainly no contrast problem here, only one of efficiency and solid angles in the detection of the neutral molecules.

DEVELOPMENTAL DIFFICULTIES

There are, of course, a number of difficulties. First, it is not possible to guarantee with certainty that fundamental results will be forthcoming as soon as suitable molecular microscopes are designed. Naturally, if the instrument does nothing but provide confirmation of known results, it will be disappointing. Secondly, although the principles are simple, the technicalities can be quite trying because of the proverbial difficulty of working with neutral molecules. Precisely the property that is their advantage for these studies make them difficult to detect, and the whole question of designing efficient ionizers with the correct properties, while maintaining a low background of irrelevant gases, is a quite taxing one. We can estimate that developing this microscope to a point where it would be useful in biological studies in a broad way is comparable to that of developing the electron microscope. By 1938, useful instruments existed, but it was more or less necessary to have an able technician, if not a Ph. D. in physics, in attendance to use the instrument. It was not until the postwar World War II period with the large expansion of electronic and vacuum sciences that packaged instruments which could be used by the essentially nontechnical in that area became available. During the period from 1928 to 1938, one can estimate that one hundred man years of engineering were expended on the electron microscope project. Facing up to this kind of task in our present research climate is not easy when many physicists say that if results can't be obtained in six months, it isn't worth doing. When one considers that there is no sharply defined problem at present well known to all biologists that can be solved by the molecular microscope, it seems like a remarkable act of faith to undertake its development at all, especially when biologists are saying that they are really over-instrumented and that their problem is not more instruments, but more theories and perhaps more correlating general principles. However, we believe that such arguments do not apply to the molecular microscope which will give dynamical information to complement the beautiful structural information obtained by electron microscopy and x-ray diffraction.

ACKNOWLEDGMENTS

We have enjoyed stimulating discussions with Drs. A. Essig, S. R. Caplan, and D. F. Waugh, as well as members of the M.I.T. Molecular Beam Group.

REFERENCES

[1]   For a general reference on molecular beam techniques, see N. F. Ramsey, 1956. Molecular Beams. Oxford, England: Oxford University Press.

[2]   Weaver, J. C.and J. G. King. 1970. MIT, RLE, Quarterly Progress Report, No. 98, p. 9.

[3]   For an example of the response of water transmission of a living membrane to an neurohypophyseal hormone, see A. Leaf and R. M. Hays. 1962. J. Gen. Physiol. 45:921.

[4]   Waugh, D. F., private communication.

CHAPTER 7

THE SCANNING ELECTRON MICROSCOPE IN BIOLOGY

Lloyd Sutfin

In the past few years the scanning electron microscope (SEM) has seen increasing use in the biological sciences. The most common use has been the visualization of surfaces, which was only accomplished with difficulty before the SEM became available because of the limited depth of field of the light microscope and the requirement for the transmission of electrons without appreciable energy loss in the conventional electron microscope. This left a large void between gross anatomy, which is essentially the study of the shape and location of surfaces, and histology, which primarily deals with structures surrounded by surfaces. There exists a need in anatomy for a cataloging of surface structure similar to that which has produced the atlases of gross anatomy and histology.

Although the SEM has been used primarily in research, its value as a communications device in the teaching of morphology should not be overlooked. As specimen preparation becomes routine and the cost of microscopes decreases, (and this has been occuring) the employment of the SEM in teaching should expand its contribution to biology and medicine. For although few people have the patience or perceptual abilities to reconstruct surfaces from serial sections, hardly anyone fails to be impressed by the nature of surfaces when visualized by stereographic pairs produced in a scanning electron microscope.

In this discussion I shall refer to the electron microscope, which has made such a large contribution to biology, as the conventional electron microscope (CEM) both because its use is very commonplace and because it is strictly analogous to a light microscope fitted with a camera. We will see later that the designation transmission electron microscope is ambiguous because both SEMs and CEMs can produce transmission micrographs.

The objectives of this discussion are:

1.   To describe the SEM in a general way
2.   To point out the fundamental limitations of the instrument dictated both by instrumental design and, more frequently in biology, by the specimen
3.   To point out the various modes of use and types of information which may be obtained

The whole area of sample preparation will be avoided except for the following observations:

1.   This area is wide open and controversial at present
2.   In contrast to metallurgical applications the preparations of biological specimens is an extensive procedure (The techniques used currently are essentially those used in conventional electron microscopy.)
3.   There is a consensus of opinion that careful preparation of biological specimens is necessary. The squabble is over which method is best.

DESCRIPTION OF INSTRUMENT

A general schematic is shown in the diagram in Figure 1.

If the image is produced by optically focussing transmitted electrons

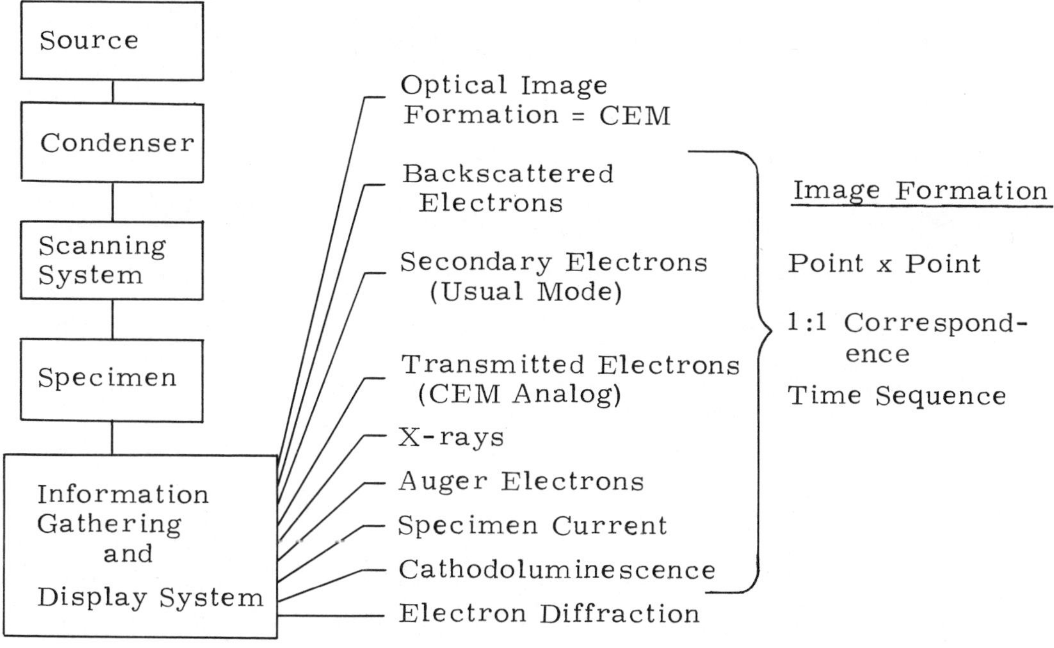

Figure 1

and the scanning system is a manually operated beam centering system
then this schematic describes a CEM. Although an optical image
formation system has not been offered as an option in a SEM, this type
of image formation has been commercially available on electron
microprobes. (Electron microprobes and scanning electron microscopes are so similar today that one is hard pressed to point out an
essential difference).

If the image is built up point by point in a time sequence then we
have a scanning microscope. Because the image depends on time
sequencing nothing must be focussed after the beam impacts the specimen and any contrast producing mechanism may be used to form an

image as long as enough signal can be collected to produce an acceptably noise-free image in a reasonable length of time.

Figure 2 shows schematically a typical commercial scanning electron microscope set up to produce an image from the secondary electrons generated at the point of impact of the beam. The primary electrons originate from an incandescent tungsten wire which has been bent to a hairpin shape. The electrons which escape from the filament are accelerated toward the anode. The filament, grid cap, and anode form an electrostatic lens which focuses the electrons producing a beam with a minimum diameter which becomes the apparent source. Proper alignment of the filament, grid cap and anode is essential to minimize the size and maximize the brightness of the source. The electromagnetic lenses further reduce the diameter of the beam. The final lens is used to bring the beam to its minimum diameter at the specimen. The beam is deflected twice by the scanning coils so that the beam pivots about a point in the final aperture and forms a square raster on the specimen. The diameter of the final aperture and the aperture to specimen distance determine the angle of convergence or aperture angle, α, of the beam. A sweep generator controls the deflection of the beam in both the display cathode ray tube and the microscope, therefore, the raster on the specimen is in synchrony with the raster on the CRT.

The interaction of the beam and the specimen produces several physical phenomena which can be used for image production. In the

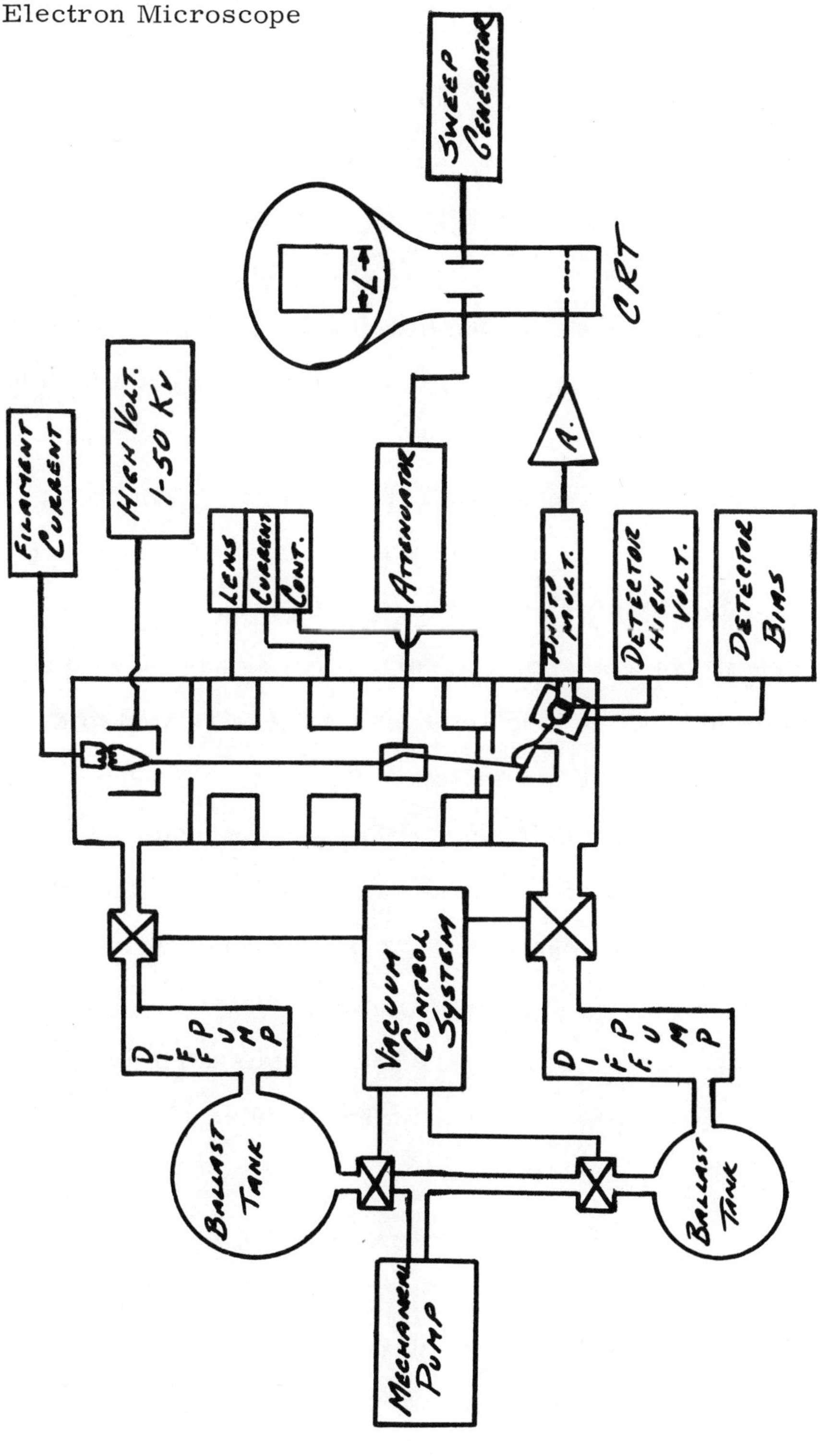
SWEEP GENERATOR
CRT
A.
FILAMENT CURRENT
HIGH VOLT. 1-50 KV
LENS CURRENT CONT.
ATTENUATOR
PHOTO MULT.
DETECTOR HIGH VOLT.
DETECTOR BIAS
VACUUM CONTROL SYSTEM
DIFF PUMP
DIFF PUMP
BALLAST TANK
BALLAST TANK
MECHANICAL PUMP

Figure 2

secondary electron mode the detector has a positive bias and the electrons are drawn to the detector; thus signal collection is very efficient in this mode. Because the detector is biased an electric field exists in the specimen chamber and may affect the specimen.

Once the electrons pass the detector grid they are strongly accelerated towards the transducer, usually a scintillator, which has a conductive coating and rides at approximately 10 KV. The light produced by the electrons is carried by a light pipe to a photomultiplier where an electrical signal is produced; this is amplified and used to modulate the intensity of the CRT beam. The result is a raster on the display tube which has a point to point correspondence with some area on the specimen and the brightness of a point is related to the strength of the signal reaching the detector.

The column and specimen chamber are evacuated. Since the specimen chamber need not be at the same pressure as the gun and column and it is sometimes desirable that the specimen chamber pressure be considerably higher, the best system is one which pumps the gun and column separately. In this type of vacuum system pressure changes within certain limits in the specimen chamber affect the gun very little.

An entirely different approach to scanning microscope design has been taken by Crewe and his associates. Since this approach is past the experimental stage and a microscope of this design is commercially available,it should be discussed. I have no doubt that the availability of this instrument will have a strong influence on scanning

microscopy both from the manufacturer's and the user's point of view.

A schematic of this instrument operating in the secondary electron

mode is shown in Figure 3. The microscope consists of a field emis-

sion gun, an aperture, a beam deflection system and a detector. The

primary virtue of this microscope is that the beam current is about

two orders of magnitude higher than that found in microscopes using

hot tungsten filaments. The importance of this will be made clear

later. The vacuum system design is more critical since pressures in

the range of 10^{-9} to 10^{-10} mmHg are required for the field emission

source. However, if the specimen chamber is also held at this

vacuum, contamination of the specimen during examination will be

less and this is an important advantage.

Field emission is caused by the potential between the tip and the

first anode and the accelerating voltage is applied to the second anode.

The configuration of the tip and anodes also provides focussing and a

beam diameter of 100Å has been obtained with the gun alone. The

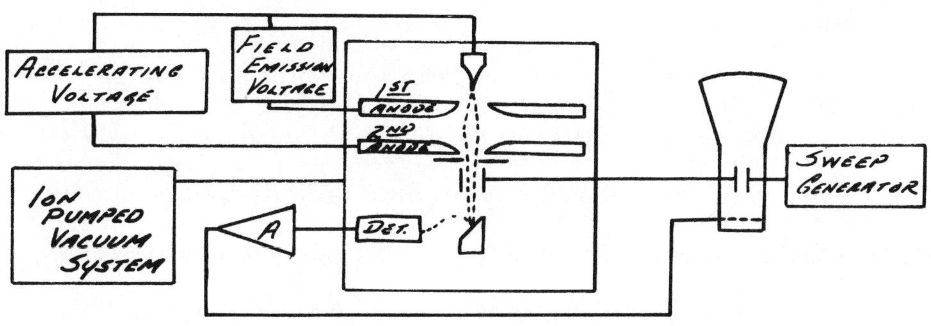

Figure 3

electrostatic deflection plates are positioned after the final aperture.

This is possible because the lens has a very long working distance.

LIMITATIONS

The limitations on the application of the scanning electron microscope

can be separated into four categories:

1. Electron Optical

    a) Beam Current
    b) Diffraction
    c) Lens abberations

2. Specimen - Beam Interaction

    a) Range of electrons in specimen
    b) Charging
    c) Radiation damage

3. Environmental

    a) Stray fields
    b) Mechanical vibration
    c) Contamination

4. Signal Collection

    a) Efficiency
    b) Sensitivity and Precision
    c) Discrimination

These limitations will be discussed from the point of view that the

scanning microscope is an instrument which can be used to provide

many types of information about a specimen besides surface morphol-

ogy. The limitation in categories 1 and 3 will apply more or less

uniformly to the several different modes of operation. Those in 2 and

4 will vary markedly depending on the phenomena being used for image

production.

Fundamental Electron Optical Relationships

Magnification (M)

Determined by the ratio of the raster size on the visual display (CRT) to that on the specimen (L and ℓ in Figure 2 and Figure 4 respectively).

$$M = \frac{L}{\ell}$$

Depth of focus (Δf)

Assume 1000 lines (N_L) raster on a 10 cm x 10 cm image. Define a picture element as a small square on the CRT image with a side, $\Delta L = \frac{L}{N_L}$. ΔL is .01 cm which is somewhat smaller than the dimension comfortably visualized by the unaided eye but it can be resolved by photographic emulsion. Assume also that the image goes out of focus when the beam diameter increases by the length of one picture element on the <u>specimen</u> ($\Delta dp = \frac{\Delta L}{M}$). Figure 4 shows these relationships.

$$\tan \alpha = \Delta dp / \Delta f$$

$$\tan \alpha = \frac{\Delta L}{M \Delta f}$$

for small α

$$\Delta f = \frac{\Delta L}{M \alpha}$$

The depth of focus depends on the electron optics and the recording conditions. Although typical depths of focus are many times greater

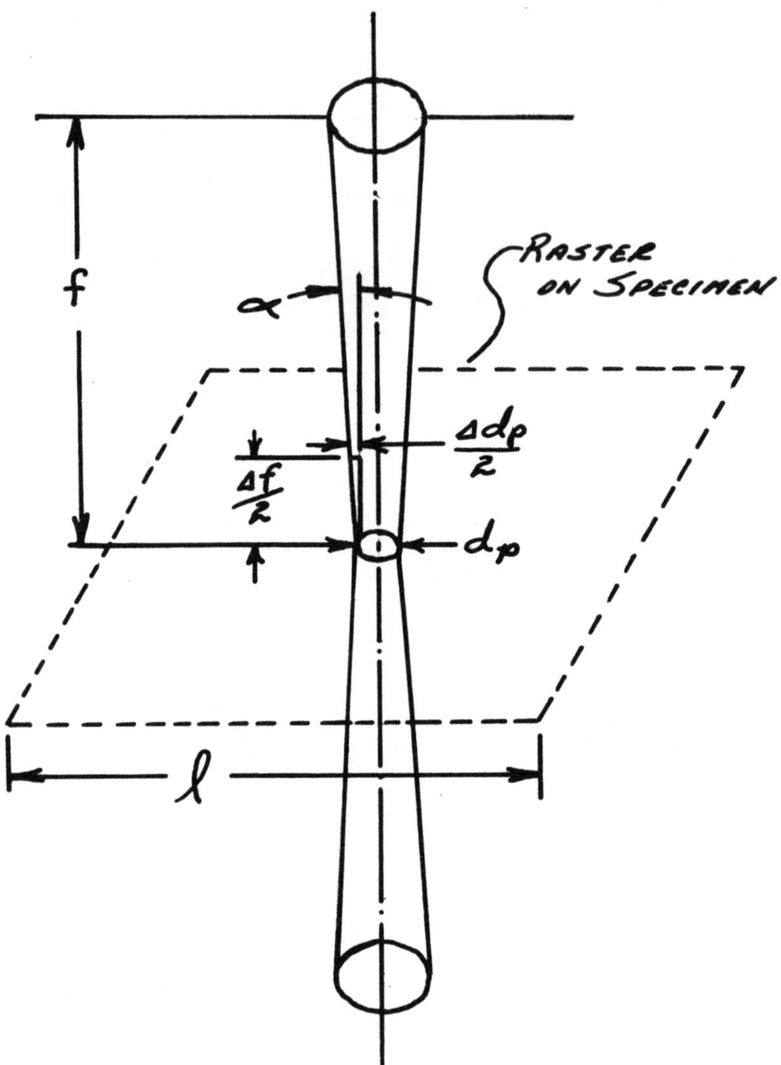

Figure 4

than those of light microscopes, they are not great enough to keep an

entire raster in focus at high magnifications at tilt angles commonly

used for contrast development in the secondary electron mode. (Usu-

ally greater than 30° from horizontal.) Auxilliary focussing coils

which automatically correct for tilt are being employed in some instruments.

Contrast and resolution

The most important property of the final image is the information
which it conveys to us about the specimen. There are two components
to visual images besides a 1:1 correspondence with the specimen,
contrast and resolution. No matter how small a feature we are able
to observe, unless we can distinguish it from its surroundings we can
learn nothing about it. In early conventional electron microscopy of
biological materials the scattering of different structures was so
similar that even though the theoretical resolution was very good,
little was observed until shadowing and staining techniques which
produced contrast were developed.

<u>Contrast</u>   In order to tell one point on a picture from another, the
signal arriving at the detector at the time corresponding to point A
must be different from that arriving at the time corresponding to
point B. Because both the current in the electron beam and the signal
generation process are random processes the signal from point A alone
will also vary. The difference in average signal from two points must
be great enough to distinguish this difference from random fluctuations
if we are to be able to detect it.

In the discussion of depth of focus we defined a picture element. Let
us assume that we wish to distinguish one picture element from another
with a minimum detectable difference, $\frac{\Delta N}{N}$. Where N is the average

number of image forming events reaching the detector in the time (t)
during which the beam is on the picture element. The number of
picture elements (N) will be: $N = (\frac{L}{\Delta L})^2$. There will also be a relationship between the events detected in any given period and the current in the beam (i_p), let

$\mu = \frac{Ne}{i_p t}$ (e is the electronic charge 1.6×10^{-19} coulombs/electron)

t is the time per picture element.

Assuming that the signal arrival at the detector is a series of
randomly spaced independent events and can be described by Poisson-
ian statistics thus:

Variance, $V = N$

Standard deviation, $\sigma = \sqrt{N}$.

Rose's criterion for distinguishing a change in signal level is:

$\Delta N \geq 5\sqrt{N}$.

If one wishes to distinguish a fractional difference (K) then:

$\frac{\Delta N}{N} = K$

$KN \geq 5\sqrt{N}$

or

$N \geq \frac{25}{K^2}$.

The magnitude of the detected signal is inversely proportional to the
square of the fractional difference in signal. N can be made as large

as one pleases by spending enough time per picture element. However,
instrumental stability and time available put real limits on N. It is
more realistic to ask how long will it take to produce an image in
which a difference of K is detectable:

$$T = Nt = \frac{25}{K^2} \frac{e\,N}{i_p \mu} \,.$$

If one assumes a 1000 line square picture ($N = 10^6$), a signal production – detection efficiency of 50 percent ($\mu = .5$) a contrast difference of 5 percent ($K = .05$) and a beam current $i_p = 10^{-11}$ amps, the
exposure time T is slightly greater than 300 seconds.

Resolution   Without sophisticated image processing, the resolution
can be no better than the diameter of the beam, dp. Langmuir's
equation relates the minimum beam diameter, d_o, which an <u>ideal</u> lens
can produce to given beam current, ip:

$$d_o^{\,2} \approx \frac{6\,ip}{\pi^2 B\alpha^2}$$

B is the brightness of the source, Amp. cm^{-2} sterad.$^{-1}$; it is a
function of the material, the accelerating voltage, V, and the temperature of the filament, T.

$B \approx J_c eV/\pi kT$.

k is Boltzman's constant, and J_c is the beam current density which is
determined by the filament material being used and the focussing properties of the gun.

The ideal diameter will be increased by diffraction and abberations according to the following:

Diffraction $d_f = \dfrac{1.22\lambda}{\alpha}$

λ = electron wave length = $\dfrac{12.3}{\sqrt{V}}$.

Spherical Abberation: (due to the fact that the focal length of the lens is dependent on the radial distance from the axis at which the electron enters the lens).

$ds = 1/2 \, C_s \alpha^3$,

C_s is the spherical aberration coefficient and is roughly equal to the focal length (f). Chromatic abberation is caused by the dependence of the focal length on the energy of the electron being focussed,

$d_c = \dfrac{\Delta V}{V} \, C_c \alpha$

C_c is the chromatic aberration coefficient and is also approximately equal to the focal length.

It is customary to assume that the contributions to the final beam size can be summed in quadrature:

$d_p^{\,2} = d_o^{\,2} + d_f^{\,2} + d_s^{\,2} + d_c^{\,2}$

In SEM work the high voltage regulation is good enough and no energy losses take place before the final lens so that chromatic abberation can usually be ignored

$d_p^{\,2} = \dfrac{6i_p}{\pi^2 B} \, \alpha^{-2} + 1.5\lambda^2 \alpha^{-2} + 1/4 \, C_s \alpha^6$

From this relationship two conclusions can be drawn.

Since two terms are inversely related to α and the third is not, there is an optimum value of α for a given system. Also if the ideal diameter, d_o, is negligible, for instance, if the brightness is very great, the diffraction and abberations produce a finite beam diameter and

$$d_p \alpha\, C_s^{1/4} \lambda^{3/4}.$$

These are the same limits which prevail in CEM. By decreasing C_s which is done by shortening the focal length ($C_s \approx f$) or increasing the accelerating voltage ($\lambda \alpha \frac{1}{\sqrt{V}}$) the resolution of the scanning electron microscope can be made to approach that of CEM. This has been demonstrated by Crewe and his colleagues for a scanning transmission microscope. Gun design improvements have allowed very small beam diameters, dp, to be obtained while beam current, i_p, has increased. The relationships between dp, i_p, and gun brightness, B, are shown qualitatively in Figure 5.

Some typical instrument characteristics are found in Table I. Thus the microscope is not limited by the scanning feature, but rather by the use to which it is put.

Table I

Cathode Type	dp (A)	f (mm)	i_p	B	Vacuum Required (mmHg)	USE
Tungsten Hairpin	≈ 50	15	8×10^{-13}	$\approx 10^5$	5×10^{-4}	S or T
Lanthanum Hexaboride	≈ 30	15	1×10^{-12}	$\approx 10^6$	10^{-5}	S or T
Tungsten Field Emission	≈ 10	0.6	1×10^{-10}	$\approx 10^8$	10^{-9}	T

S $\equiv$ surface study, T $\equiv$ transmission (after Broers, 1970)

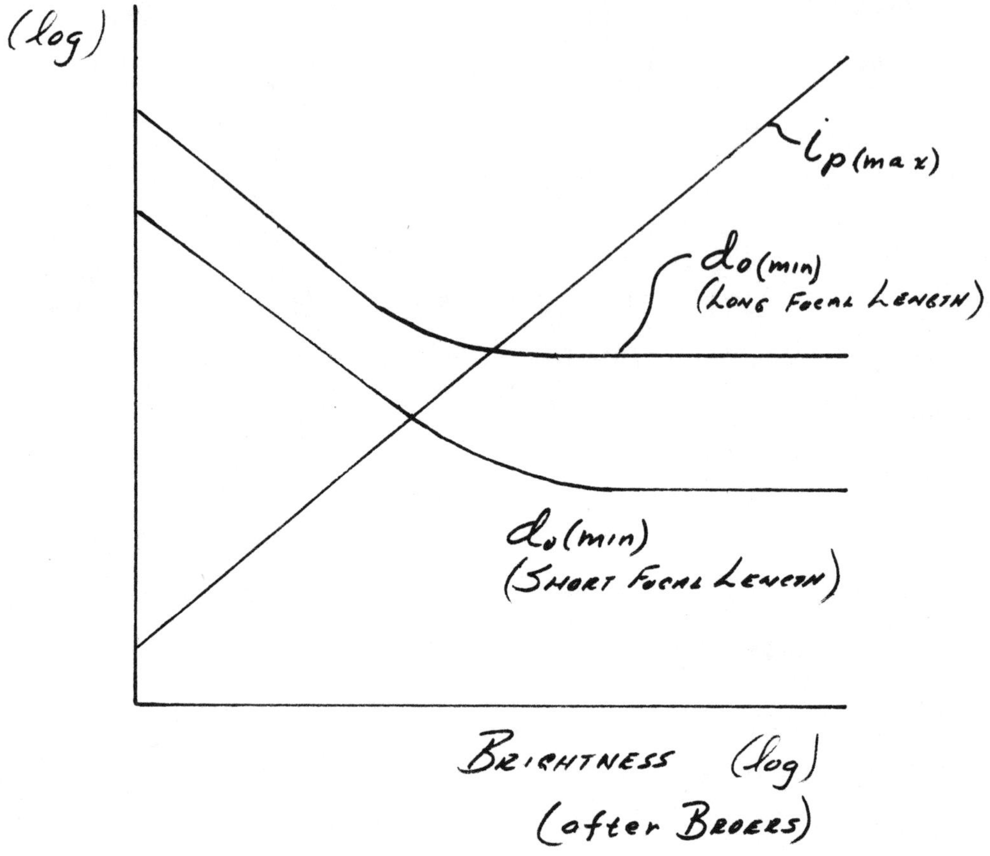

Figure 5

<u>Modes of Use</u>

Several types of signal can be used to produce an image. They can be

classified as follows:

I.   Electrons

   1.   Escaping from surface

      a.   Backscattered
      b.   Secondary
      c.   Auger

   2.   Transmitted

      a.   Unscattered
      b.   Scattered
            Elastic
            Inelastic

   3.   Conducted (Specimen Current)

   4.   Reflected (Mirror)

II.   Photons

   1.   X-rays

   2.   Light (Cathodeluminescence)

Because of the easy access to the specimen often several different types of detector can be used either simultaneously or sequentially without a long conversion procedure. The absence of lenses after the specimen allows the detector to be placed quite close to the specimen and the interception of radiant signals can be relatively efficient.

A brief description of each type of signal will be given and then a discussion of the limitations imposed by the specimen-beam interaction will follow. The electrons coming from the surface of a bulk specimen will have an energy distribution as shown in Figure 6. The electrons which have energies nearly equal to that of those in the beam are called backscattered electrons: those with energies of a few electron volts are called secondary electrons. The fraction of backscattered electrons increases as the average atomic number of the specimen increases, thus backscattered images can yield information regarding the chemistry of the specimen. There is a relationship between composition and secondary electron production but it is not well understood

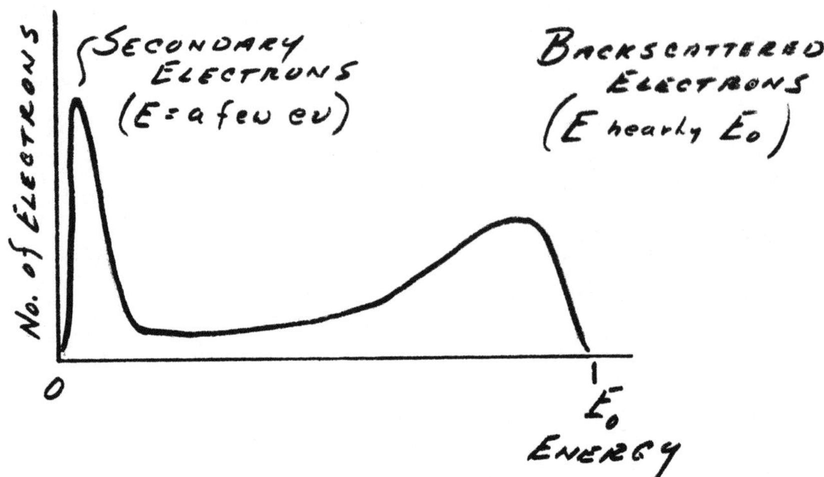

Figure 6

and generally speaking secondaries are used more for topography than chemistry.

The energy distribution of backscattered and secondary electrons is continuous. There are electrons which will have discrete energies. One type, Auger electrons, can be used to give information on the chemistry of the specimen. These electrons are produced when atoms undergo innershell excitations (as in X-ray production). If the energy of the electrons coming from the specimen is analyzed, peaks will be found at energies corresponding to electronic energy levels of the elements present in the target. The energy resolution of this type of analysis is quite good (a few electron volts) and it holds promise for chemical analysis in the light element region of the periodic table.

When the specimen is a thin section the transmitted electrons can be used to produce an image. Many electrons will not undergo scattering at all. If these are detected and the signal used to modulate the display CRT, a bright field image will be produced. That is, areas of high density, which scatter electrons more, will be dark while regions which are less dense will appear brighter. If scattered electrons are detected, then a dark field image will be produced which generally will be the reverse, dark for light, of the bright field image.

Scattering of the electrons is of two types - elastic (no energy loss) and inelastic (energy loss). Crewe has used the ratio of elastic to inelastic scattered electrons to produce images of thin specimens. This ratio is proportional to the atomic number of the scatterer and yields information regarding the chemistry of the target. Using concentric detectors and energy analyzer and an electronic circuit which produces a signal proportional to the ratio to modulate the CRT intensity a visual image can be produced as the specimen is scanned.

If an electron energy analyzer is placed so that it intercepts electrons coming from the specimen, images can be produced with electrons which have a specific energy. Electrons which have undergone a selected energy loss can be brought to focus on a detector while all others are excluded. The image produced has superior contrast to that produced with unfiltered electrons - the energy analyzer serves the same function as an aperture. The energy loss spectrum also contains information about the electronic structure of the specimen and can yield information regarding its composition.

Auger electron can also be detected from a thin specimen. If the specimen is placed in a co-axial energy analyzer the spectrum of energies of electrons leaving the bottom of the specimen can be obtained. Here again peaks at energy levels corresponding to those excited in the specimen will be found.

A fraction of the electrons will remain within the specimen. If the specimen is grounded through a current measuring device, this signal can be used to form an image. The magnitude of the specimen current is the difference between the beam current and the backscattered and secondary electron currents.

Although backscattered electrons are often referred to as being reflected from the specimen, a specific mode called the mirror mode exists. It is possible to bias the specimen to the same potential as the acceleration voltage and the electrons will not be energetic enough to reach the specimen. If the bias is slightly greater than the accelerating potential, they will be accelerated away from the specimen. Electron detectors placed above the specimen will produce a signal which is dependent on the bias at the surface of the specimen. This is the principle of the mirror microscope, and it provides an instrument for studying voltage variations without inducing currents in the specimen with the electron beam.

Electrons lose discrete amounts of energy by exciting the electronic structure of the specimen. Excitation (creating a vacancy) of the innershell electrons of an atom produces a characteristic X-ray spectrum. This spectrum can be used to identify the elements present and if the intensity of a single characteristic X-ray line is used to modulate the CRT intensity a composition map (X-ray image) is produced. Excitation of chemical bonds may produce light, hence cathodoluminescent spectra can yield information on the molecular species present. Chemical analysis by X-ray spectra is fairly well developed especially for relatively simple metallurgical and semiconductor systems.   Chemical analysis by cathodoluminescence has only begun to receive serious attention.

SPECIMEN-BEAM INTERACTION

If the phenomena which we have listed as being useful contrast producing mechanisms took place within a region with dimensions comparable to that of the beam diameter, scanning electron microscopy would be much less complicated. However, the electrons after impact with the specimen penetrate through the specimen for a distance depending on the density of the specimen and the energy of the electrons in the beam. They move into the specimen and spread laterally and the diameter of the region being excited can be quite large compared to the beam. For example, the range of 30 kev electrons in a metal such as copper is several microns and frequently the resolution of a scanning electron micrograph is determined by the range of the electrons in the specimen rather than the diameter of the beam. Figure 7 diagrams roughly what the envelope of electron ranges may look like under different

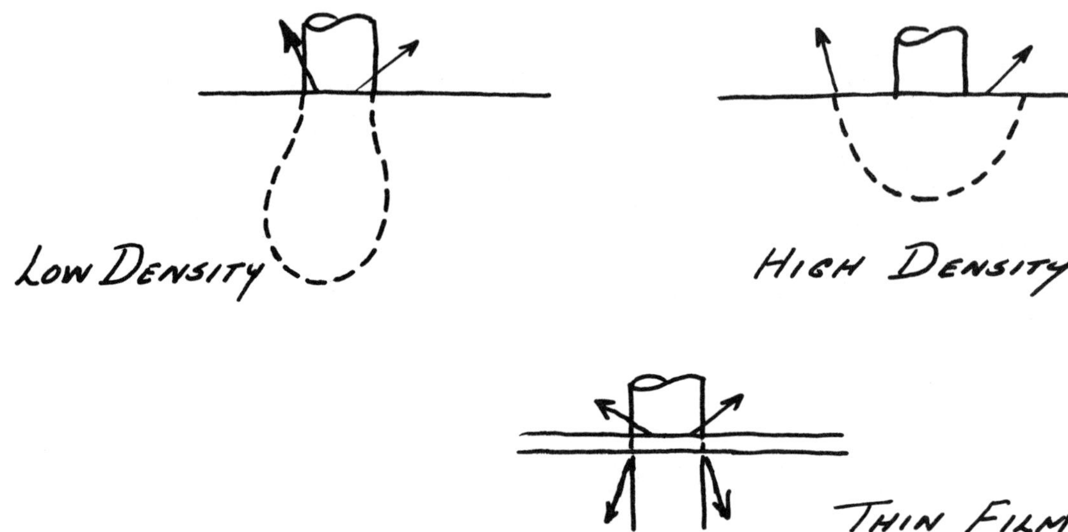

Figure 7

conditions.

The resolution of X-ray analysis is the least dependent on beam

diameter or the one which is most limited by specimen-beam inter-

action. Although backscattered electrons emerging from the specimen

lateral to the beam produce secondary electrons, a large fraction of

the secondaries are believed to come from the area of impact and the

fraction that comes from the lateral portion apparently just increases

the background signal level and is not an important detriment to image

formation.   Therefore, the resolution of secondary electron scanning

micrographs closely approaches the beam diameter.   Figure 8 is a

sketch showing the separate contributions of the beam and backscatter-

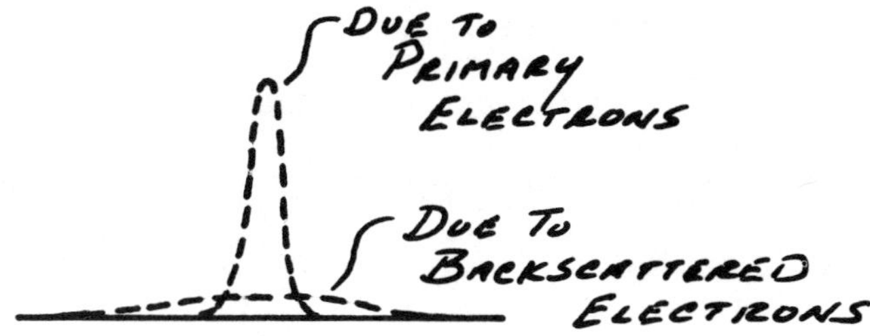

Figure 8

ed electrons.

The resolution of backscattered electron images is poorer than that made with secondary electrons for two reasons. The backscattered electrons are more uniformly distributed over the excited region than secondaries and the fraction of all backscattered electrons which is collected is smaller since they are not "focussed" by the grid of the detector, which may have a negative value in this mode.

Although Auger electrons are produced by the same excitation which produces X-rays, electrons do not have the penetrating ability of X-ray photons and can undergo fractional energy losses. The distance which an Auger electron can travel within the specimen without being absorbed or having lost so much energy as to be unrecognizable is only a few angstroms. Auger analysis is considered to be a surface analysis because of this limit in depth. The lateral spread of the primary electrons enlarges the area of origin of Auger electrons.

One approach which has been taken to improve the spatial resolution of scanning microscopy is to prevent the lateral spreading and limit the size of the excited volume by analyzing thin sections. This is particularly appliable to biological studies since thin sections are readily prepared. One difficulty with this technique is that the signal level is low and very efficient detection methods are required. Signal levels may be so low that image proudction takes too long and point analysis must be used. This is especially true of X-ray and Auger analysis in microscopes using hot filaments. Improvements are expected as field emission guns become available.

One consequence of bombarding a non-conducting specimen with electron is charging of the specimen. Charging may increase secondary electron production and distort or deflect the beam. Biological specimens can be treated with anti-static agents or coated with conductive material such as carbon, aluminum, or gold. Even though the coating may be continuous over the surface (this is not always attained in practice) sub-surface accumulation of charge may still occur.

Secondary electron micrographs of biological specimen often have an area of very high brightness which detracts from the esthetic qualities of the micrograph and which reveal little in that area since the system is saturated and no contrast exisits. Two instrumental variables can be changed to improve the information obtained from such specimens. An improved image is very often obtained if the kilovoltage is decreased and if the detector bias is reduced. This is a prime example of the specimen rather than the electron optical condi-

tion limiting the performance and a compromise must be made to
improve the result. SEM operators seem to be too reluctant to change
microscope operating conditions from those which are said to produce
best results on ideal specimens. Decreasing the accelerating voltage
increases the secondary electron yield uniformly over the specimen
and decreases the penetration of electrons below the surface, both
tend to reduce charging. Decreasing the detector bias will reduce the
field in the region of sharp edges. It is felt that this field increases
electron emission from the specimen producing "bright" areas at
edges. Regardless of the correctness of these assumptions both
changes produce a better micrograph of non-conducting specimens.

ENVIRONMENTAL LIMITATIONS

The concluding portion of this discussion will deal with what may be
the most serious problem in scanning microscopy at present - contamination. Although stray fields and vibration may distort the image,
the instrument can usually be sufficiently shielded and isolated to
eliminate their effect, once recognized. Contamination is extremely
difficult to avoid since the source of contaminants may be the specimen surface, the operator's technique in handling the specimen and
specimen chamber of the microscope, or the vacuum system itself.
It is generally considered that the contamination layer which builds up
on the specimen surface is due to the polymerization of organic molecules in the area of specimen-beam interaction and the layer was

considered to be carbon. However, X-ray microanalysis has recently
shown that an important contaminant is silicon. Silicon containing
vacuum fluids and greases should be avoided but even where they have
never been used, silicon contamination has been found. Although not
proven, an apparent source of silicon is semiconductor devices. The
study of these devices is at present the economic backbone of scanning
microscopy, and it is difficult to find an instrument which has not been
used to study them. Apparently enough radiation damage occurs to
contaminate the vacuum system and cause the subsequent deposition
of silicon on other types of specimen. This not only adds an unwanted
characteristic X-ray signal (or Auger), it also increases the density
of the volume being analyzed so that quantitative analysis is not possible.

Two approaches to eliminating the vacuum system's contribution to
this problem are at hand. Ultra-high vacuum such as obtained with ion
pumps may eliminate it sufficiently if extraneous sources of contamination can also be eliminated. We have found that directing a jet of
inert gas onto the surfaces of the specimen reduces contamination to
the point where it is negligible even with a rather unsophisticated,
"dirty", vacuum system and with no more than normal precautions as
far as handling the specimen. This system has been shown to clean
the surface of metal specimens under electron bombardment.

HIGH SPATIAL RESOLUTION X-RAY MICROANALYSIS

One biological application for which the SEM is well suited is the microanalysis of thin sections of biological tissue. Although the visualization of subcellular structures with the CEM has reached a high level of sophistication, direct analysis of the chemical composition of these structures has not been possible. Electron probe microanalysers have not proved suitable for this purpose because of restrictions placed on the strength of the x-ray signal, beam diameter, and magnitude of the excited volume by the sensitivity of diffracting x-ray spectrometers. The advent of semiconductor x-ray detectors with efficiencies several orders of magnitude better than diffracting spectrometers, has removed some of these limitations so that analysis of structures with diameters of several hundred angstroms is possible with significant improvements to be expected.

Figure 7 shows that the only way the volume analyzed can be reduced to the submicron level is to decrease the thickness of the specimen. For 20 kev electrons and 1000Å thick sections, it is adequate, although not strictly correct, to visualize the excited volume as being a truncated cone with an upper diameter equal to that of the beam and the lower diameter equal to the thickness of the specimen. The excited volume is in the range of 10^{-3} to 10^{-4} cubic microns and this is small enough to analyze selected ultrastructural features of cells. To adequately visualize the specimen, a transmitted electron detector is required. Manual positioning of a stationary electron beam allows the probe to be accurately placed.

In applications to date, contamination has been adequately reduced by gas jet decontamination to allow qualitative analysis for the presence of calcium and phosphorus in granules of mitochondria with diameters ranging from 500 to 1000Å with a standard SEM. Improvements in electron optics, x-ray detector resolution, electron velocity analyzers, and vacuum technology as applied to SEM's will make this type of application very important to the future of the SEM in biology.

CONCLUSION

The scanning electron microscope can be used to obtain several different types of information about biological specimens. Although the secondary electron mode is the most widely used to date, this mode will probably be of secondary importance in the future. Secondary electron micrographs will be more important in the teaching of anatomy, whereas the transmission mode will probably be more important to biological research. The ease with which quantitative electron microscopy can be performed in the scanning transmission mode will extend the scope of electron microscopy from being mostly descriptive to a highly analytical science. Besides the improvement in instrumentation, such as field emission guns, clean ultrahigh vacuums, and various types of detectors, specimen preparation techniques which preserve not only the anatomy, but also the chemistry of the cell, must be developed as well as proper methods of

processing and interpreting the data obtained. The tasks involved

will not be easy, but should provide a fruitful field for investigators

who are oriented in both physics and biology.

GENERAL BIBLIOGRAPHY

Oatley, C.W. , W. C. Nixon, R. F. W. Pease. 1965. Scanning Electron Microscopy Adv. Electronic Electron Phys. 21:181.

Hayes, T. L. , R. F. W. Pease. 1968. The Scanning Electron Microscope: Principles and Applications in Biology and Medicine. Adv. Biol. Med. Phys. 12:85.

Crewe, A. V. 1970. The Current State of High Resolution Scanning Electron Microscopy. Quarterly Rev. of Biophys. V3, N1, pp. 137.

Note: Each year an excellent symposium on scanning electron microscopy is held at the Illinois Institute of Technology. The proceedings of this symposium are an excellent source of information regarding current techniques and applications in this field. For simplicity a few papers will be cited, grouped by the year in which they appeared in the proceedings. An extensive bibliography is also published in these proceedings each year.

Broers, A. N. 1970. Factors Affecting Resolution in the S. E. M. , p. 1.

MacDonald, N.C., H. L. Marcus, P. W. Palmberg. 1970. Microscopic Auger Electron Analysis of Fracture Surfaces, pp. 25.

Sutfin, L. V. , R. E. Ogilvie. 1970. A Comparison of X-ray Analysis Techniques Available for Scanning Electron Microscopes, p. 17.

Swift, J. A. and A. C. Brown. 1970. Transmission Scanning Electron Microscopy of Sectioned Biological Materials, p. 113.

Brandis, E. K. , F. W. Anderson, and P. Hoover. 1971. Reduction of Carbon Contamination in the SEM, p. 505.

MacDonald, N. C. 1971. Auger Electron Spectroscopy for Scanning Microscopy, p. 89.

Pease, R. F. W. Fundmanetals of Scanning Electron Micrsocopy, p. 1.

Sutfin, L. V., M. E. Holtrop, R. E. Ogilvie. 1971. Microanalysis of Individual Mitochondrial Granules with Diameters Less than 1000 Angstroms. Science. 174:947.

Yew, N. C. 1971. Dynamic Focussing Technique for Tilted Samples in Scanning Electron Microscopy, p. 33.

Sutfin, L. V. 1972. High Spatial Resolution X-ray Microanalysis of Thin Specimens in the Scanning Electron Microscope. In Scanning Electron Microscopy, Johari, O. and Corvin, I., eds., p. 65. Chicago: ITTRI.

PART III

MEDICAL PROBLEMS AND THE APPROACHES PROVIDED

BY MATERIALS SCIENCE

CHAPTER 8

JOINT WEAR

Eric L. Radin

All of us have had the experience of some close friend or relation suffering from arthritis. There are two distinct classes of arthritis. One is basically an inflammatory type and includes such diseases as rheumatoid arthritis, rheumatic fever, gout, and spondylitis. The other, and far and away the most common, is called degenerative or osteoarthritis and because its incidence is age related, it was felt for many years that it was simply a wear and tear phenomenon. After all, what other bearing do we know of that lasts for 65 or more years under almost constant load without maintenance and operates at coefficients of friction that are in the neighborhood of 005. It seems only reasonable that such bearings would eventually wear.

For some years now, a collaborative effort has been underway between the Department of Mechanical Engineering at M. I. T. and the Orthopedic Research Laboratories at Harvard Medical School to look at the relationship between the mechanical, structural and chemical aspects of degenerative joint disease. This work has since developed into a materials problem and the addition of the Department of Metallurgy and Material Sciences at M. I. T. to the collaboration has been found to be necessary. I will attempt here to briefly summarize our current feelings about joint wear.

Figure 1 is a diagramatic picture of what constitutes a freely movable animal joint. The bone ends are covered by a bearing surface, articular cartilage. The joint space is surrounded by a joint capsule, the inner lining of which is called the synovial membrane and this secretes synovial fluid which is the joint lubricant. The shafts of the bone are made up of rather hard bone, compact in its structure. The ends of the bone directly under the joints are composed of cancellous or spongy bone which is quite cellular in its gross appearance on cut section (I use "cellular" here in its engineering sense, meaning that there are multiple holes or chambers in the structure). In life these holes are filled with a marrow, fat, and other cellular and liquid elements.

For the last 20 years, the majority of research done in joint degen-

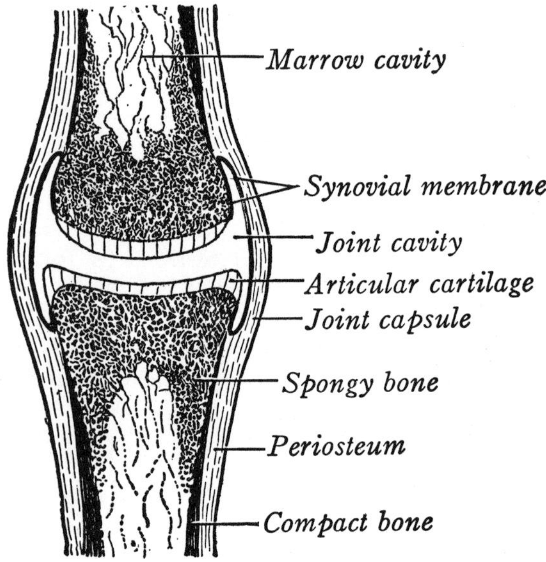

Figure 1

eration has been aimed at elucidating the metabolic defect that causes cartilage to lose its integrity and wear. A few years ago, one of the leading cartilage biochemists admitted that no such underlying defect could be found [1:152]. He proposed instead that the enzymatic degradation of cartilage is set off by some increase in the operational stress in the system.

The animal joint is lubricated with very low coefficients of friction so that even under a load of 400 pounds, the frictional force in a typical joint would only be about 2 pounds. Most of the forces that the joint is subjected to are the result of compression (Figure 2). Body weight is only part of this compressive load. Actually most of it is generated by the muscles which span the joints and contract in order to give angular acceleration to the limb or maintain balance to keep us in the upright position. Theoretical calculations of the loads in the hip and knee [2:607][3:431] have been confirmed by actual measurements [4:88] to show that actual forces in joints are multiples of the body weight, conservatively 3 or 4 times the body weight so that the compressive forces are really considerable. If this bearing were to wear out, it would make sense that it would be the compressive forces, which would be greater, which would cause it to wear.

How does the joint handle longitudinal stress? In actual daily living, the kinds of high stress most commonly applied to joints are those of impact loading: walking, running, getting up from a chair. These all involve loading cycles that have very short duration peaks. Even

FORCES ACTING ON JOINTS

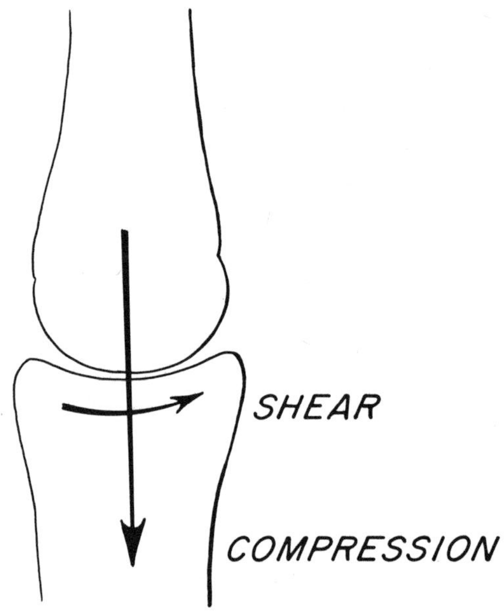

**If the load is 400 pounds,
the frictional force will be 2 lbs.**

Figure 2

grasping to pick up something is a kind of impulse activity in terms of

the joint loading. Also, if a high stress is potentiated for any length of

time, the underlying bone will break. We therefore set about to study

how the joint handles impact loading. A rig was devised which simply

dropped a weight on a whole joint. The peak transmitted dynamic force

was recorded (Figure 3). The joint was then stripped sequentially of

its tissues. First the capsule and the ligaments were cut, then the

joint lubricant, synovial fluid, was removed and replaced with a buffer,

and finally the cartilage itself was removed leaving just bare bone. It

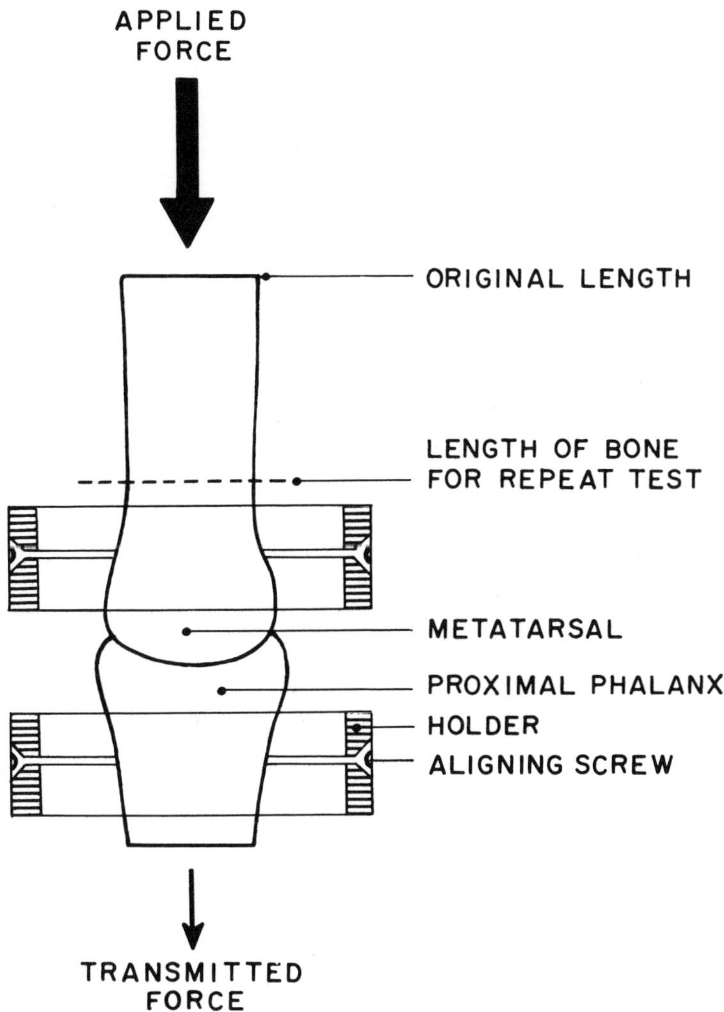

Figure 3

was found that the peak dynamic forces transmitted were increased

tremendously when the capsule was cut, but that the removal of syno-

vial fluid and the removal of cartilage made no difference (Figure 4)

[5:139].

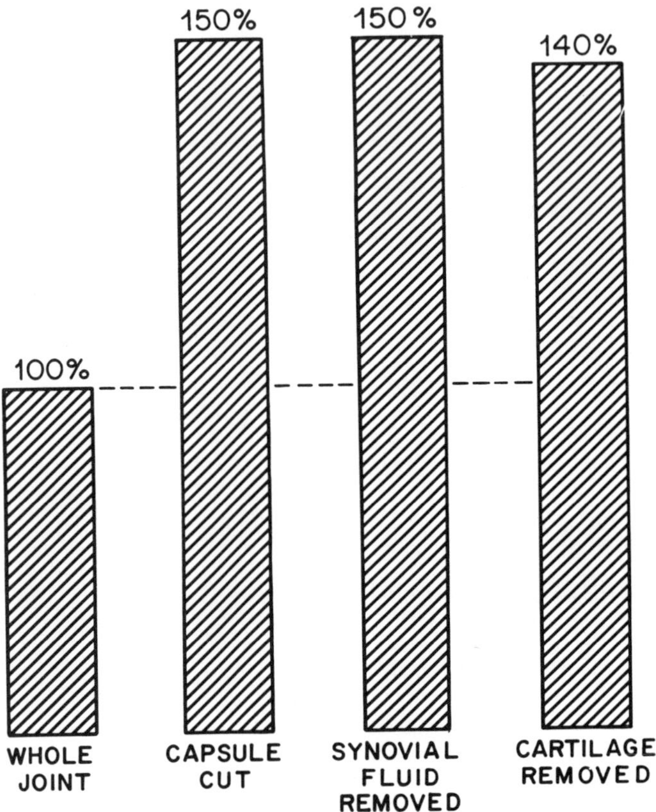

Figure 4

The next experiment then was the use of a half joint and a curved plunger with a LVTD parallel to the system so that we could measure deflection as well as peak transmitted force. In this system it was possible to remove substantial portions of the bone (Figure 5). We found that although removal of the cartilage made no difference, re-

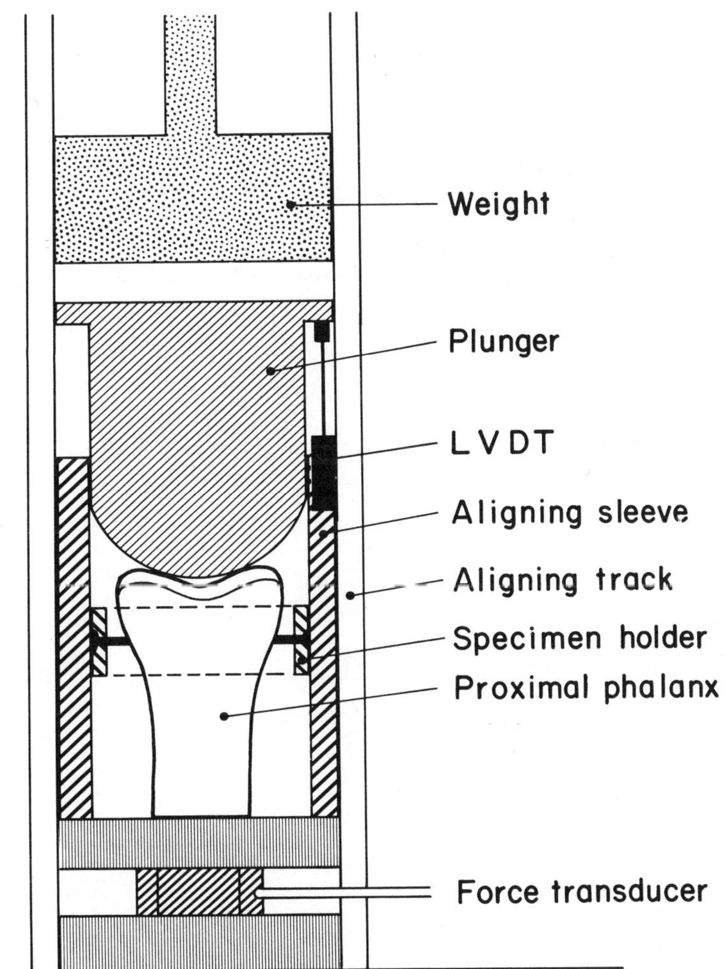

Figure 5

moval of the bone made a tremendous difference (Figure 6) [6]. These experiments were run in parallel with examination of the relative stiffness or modulus of elasticity of bone and cartilage plugs in a plug holder put into the same weight dropping device. We found that under applied stresses within the physiological range, that cancellous bone was about 10 times stiffer than articular cartilage (Figure 7) [7:444].

IMPACTION DATA
(HALF   JOINT)

	INITIAL PEAK FORCE (pounds)	MAXIMUM DEFLECTION (inches x 10^{-2})
INITIAL	190	0.164
MINUS CARTILAGE	200	0.145
MINUS 48% OF BONE LENGTH	300	0.091

(2.27 pounds dropped $\frac{1}{16}$ inch)

Figure 6.

The maximum thickness of articular cartilage in the human body is 6 mm. The maximum thickness of the cancellous bone in the human body is a few inches. So it is quite obvious that there is enough bone to do the job but not enough cartilage. The synovial fluid thickness is approximately 1000 angstroms. Since force attenuation and energy absorption require deformation, there is just not enough synovial fluid and articular cartilage to meaningfully deform. Thus it appeared as if the bone, especially this cellular cancellous stuff directly under the cartilage, was what was responsible for cushioning the shock that the articular cartilage might feel.

STIFFNESS OR MODULUS OF ELASTICITY (kg/cm^2)

APPLIED STRESS (kg / cm^2)	10	20	30	40	50
CARTILAGE	220-250	400-420	500-600	650-700	750-800
BONE	6500	7500	7800	7800	7800

Figure 7

One could then hypothesize that stiffening of this bony layer would diminish its shock absorbing capabilities and allow much higher stresses to be felt by the articular cartilage. We therefore looked at the cancellous bone directly under the cartilage in patients who had died and were autopsied and as an incidental finding had mild arthritic changes in their knees which they were unaware of. We rated the degree of arthritic change histochemically and graded it as normal, very slightly, early, medium, and obvious early osteoarthritis. Figure 8 is a bar graph showing the results. The ordinate is relative energy absorption. You can see that the normals have a certain energy absorption. The specimens from patients with very early signs of arthritis had considerably less energy absorption and in the later two

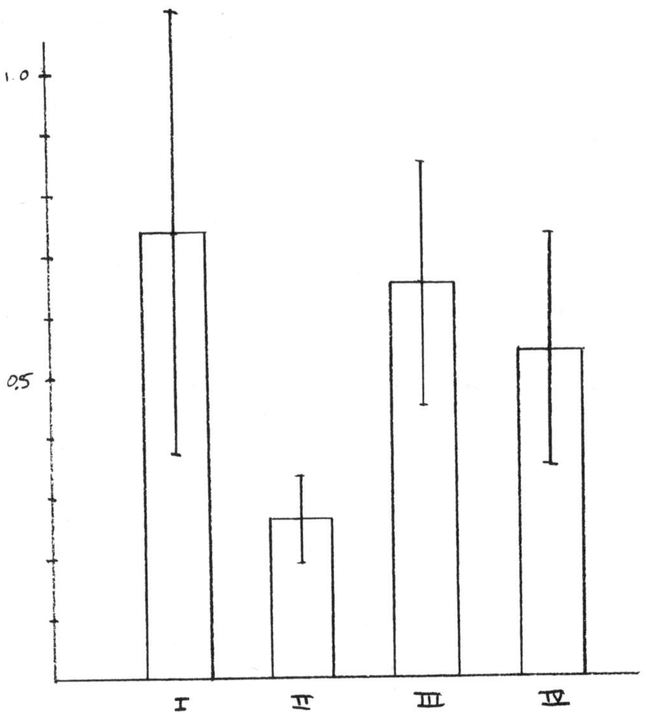

Figure 8

gradings the energy absorption went back into the normal range. Their

standard deviations are plotted on these graphs.   The differences in

the early arthritic specimens was quite statistically significant [8:400].

It is theoretically possible for bone stiffness to increase the stress

on the cartilage and in fact patients with early arthritis have such

stiffness.   The question then is how susceptible is the cartilage to

impact loading.   We then designed an experimental rig which oscillated

the joint under load.   The rig was designed so that we could measure

the instaneous coefficients of friction.   The rig was then fitted with a

magnetic clutch so that we could stop the oscillations in mid-cycle and

furthermore, the rig was fitted with a pneumatic loading device which could be activated so that when the joint was stopped in mid-cycle, an extra impact load could be applied (Figure 9). Figure 10 presents the coefficients of friction over time under three separate sets of conditions. The upper Figure A is the coefficient of friction of a joint run for 500 hours oscillated under 1000 pounds of load. You can see that the coefficient of friction even at 500 hours remains unchanged. Looked at microscopically, these joints showed no obvious signs of wear. The middle plot (B) is the result from a joint oscillated under 1000 pounds of load for 200 hours with a 2 second interruption in the midst of the cycle. Microscopically the joint resembled the first joint and there was no increase in the coefficient of friction over this period of time. The lowest plot (C) represents a joint that was oscillated under the same conditions as the middle joint, in other words with a 2 second

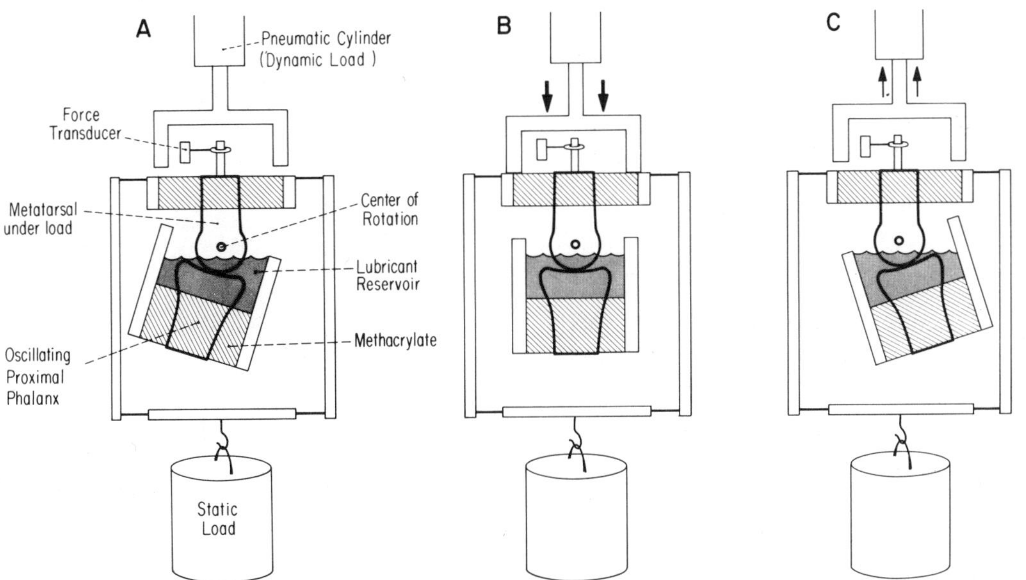

Figure 9

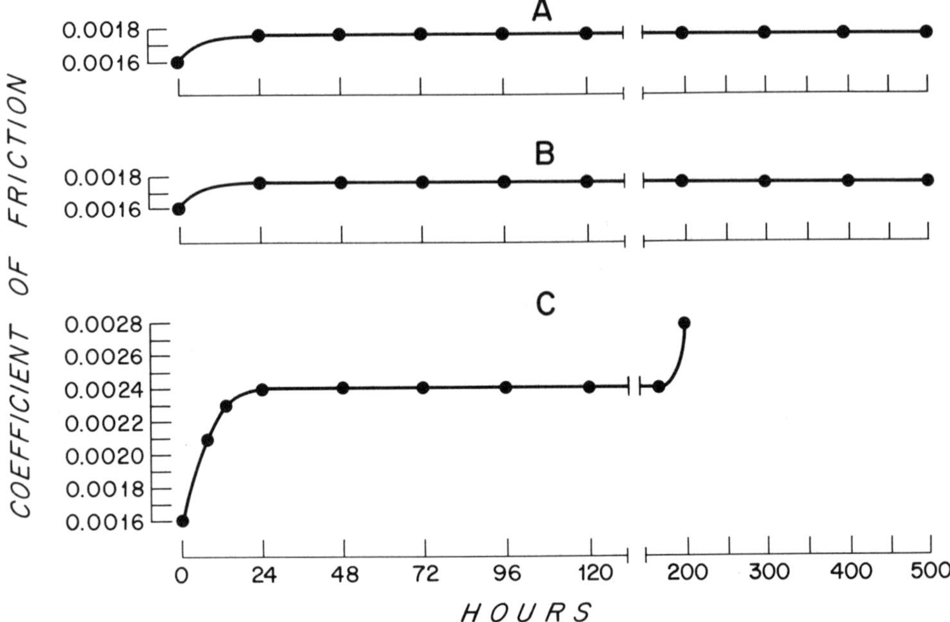

Figure 10

stop, but instead of having 1000 pounds of static load throughout the cycle as did joints A and B, the joints in group C had only 500 pounds of static load and had 500 pounds of dynamic load as applied through the pneumatic cylinder in mid-cycle. You can see the coefficient of friction rapidly rises and at the end of 120 hours, the joint was completely worn down. We concluded that articular cartilage is quite resistant to sheer stress but quite sensitive to longitudinal impact loading [9].

If there is indeed a relationship between impact loading, stiffening of subchondral bone, and degenerative joint disease, one should be able to, by applying impact loading to live joints, create all these changes. We took guinea pigs and suspended them from a rack above an impacting table. The hind legs were fitted with splints so that both legs would be held in exactly the same position. The splints on the right touched the impacting table and transmitted its force up to the knee. The splint on the left did not touch the impacting table (Figure 11). The guinea pigs were subjected to impact loading for a short period once a day. Otherwise the splints were removed and the animals were allowed to run free in their cages. We studied three parameters with time. The first one was gross appearance of the cartilage of the joints. The second, histochemical appearance of the mucopolysaccharide of the cartilage, an excellent indication of cartilage health. The third was the relative stiffness of the bone. It can be seen from the next figure (12) that the bone stiffened after just a few days of impact loading and that this was followed by changes in the cartilage. The joints then went on over some weeks to severe degenerative arthritis [10].

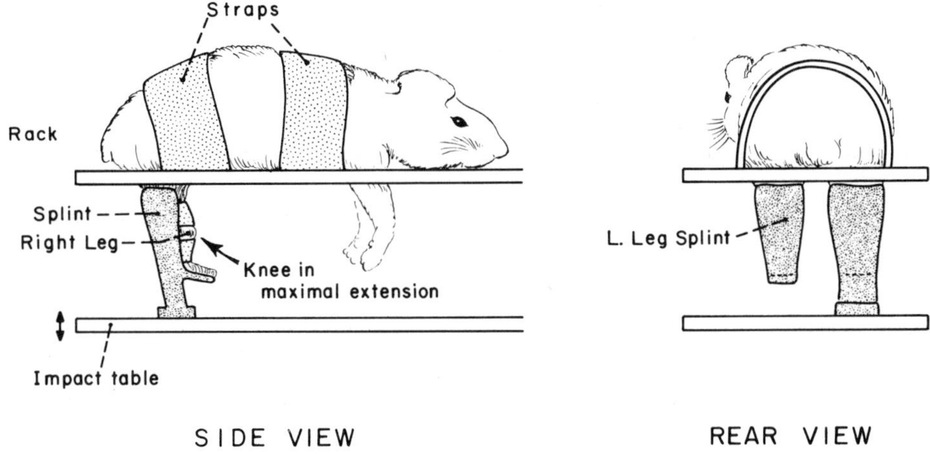

Figure 11

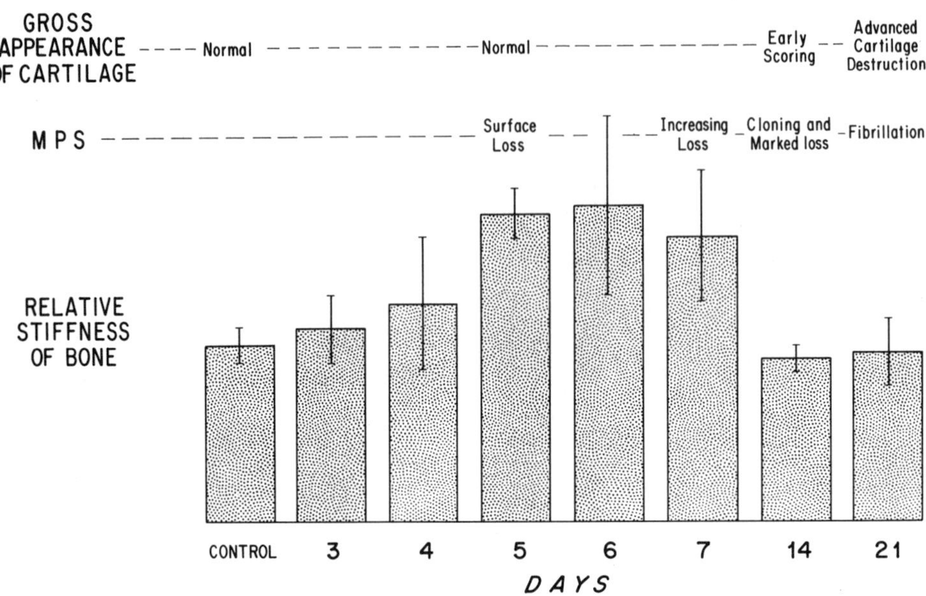

Figure 12

Our concept of what causes degenerative arthritis is the following:
Repetitive impact loading over the course of some time leads to bone
remodelling. It has been known for many years [11:389] that living
bone will respond to stress by laying down new bone. It has been
shown, more recently, that this stress must be intermittent [12:252].
This results in stiffening of the bone and because the shock absorbing
cushion is lost or diminished, there is an increased stress on the
cartilage, the cartilage then breaks down and this leads to joint degeneration.

The nature of the bony changes which diminish the shock absorbing
properties of bone may well be a microfracturing of the cancellous
bone with resultant healing. In the healed state with excessive scaf-

folding around the fracture, which the cells have put there as part of the healing process, the bone is going to be stiffer. The stiffness would diminish when the bone remodels and returns to its natural state. We have observed such changes [13]. It is possible that there are also vascular or other causes for the change in the mechanical properties of the bone. These have yet to be fully explored.

For a long time it has been felt that mechanical stress really was not to be considered in joint degeneration because many people did hard work all of their lives and had no joint degeneration. It could not be shown that articular cartilage aged significantly either chemically or mechanically [14]. What the work we have presented here shows is that indeed mechanical forces are playing a role in joint degeneration; it is just that their nature is of a slightly different type than it was assumed had been functioning. It is hoped now that further work in this area will lead to preventative measures to alleviate future suffering and disability from this most common condition.

REFERENCES

[1] Bollet, A. J. 1969. An essay on the biology of osteoarthritis. Arth. and Rheum. 12:152-163.

[2] Inman, V. T. 1949. Functional aspects of the abductor muscles of the hip. J. Bone Joint Surg. 29:607-619.

[3] Morrison, J. B. 1970. The mechanics of muscle function in locomotion. J. Biomech. 3:431-451.

[4] Rydell, N. W. 1966. Forces acting on the femoral head prosthesis. A study on strain gague supplied prosthesis in living persons.

Acta Orthop. Scand. Suppl. 88.

[5] Radin, E. L. and I. L. Paul. 1970. Does cartilage compliance reduce skeletal impact loads? The relative force attenuating properties of articular cartilage, synovial fluid, peri-articular soft tissues and bone. Arth. and Rheum. 13:139-144.

[6] ___. 1971. Importance of bone in sparing articular cartilage from impact. Clin. Orthop. 78:342-362.

[7] Radin, E. L., I. L. Paul, and M. Lowy. 1970. A comparison of the force transmitting properties of subchondral bone and articular cartilage. J. Bone Joint Surg. 52A:444-456.

[8] Radin, E. L., I. L. Paul, and M. J. Tolkoff. 1970. Subchondral bone changes in patients with early degenerative joint disease. Arth. and Rheum. 13:400-405.

[9] Radin. E. L. and I. L. Paul. 1971. The response of joints to impact loading: I. In vitro wear. Arth. and Rhuem. 14:356-362.

[10] Simon, S. R., E. L. Radin, and I. L. Paul. 1971. The response of joints to impact loading: II. In vivo wear. J. Biomech. (In press).

[11] Wolff, J. 1870. Ueberdie Innere Architektur der Knocken und ihre Bedentung fur die Frage von Knockerwachstrum. Virch Arch 50: 389.

[12] Bassett, C. L. A. 1968. Biologic significance of piezoelectricity. Calc. Tiss. Res. 1:252-272.

[13] Radin, E. L., H. G. Parker, J. W. Pugh, R. S. Steinberg, I. L. Paul, and R. M. Rose. Response of joints to impact loading: III. Relationship between trabecular microfracture and cartilage degeneration. Submitted for publication.

[14] Sokoloff, L. 1969. The biology of degenerative joint disease. Univ. of Chicago Press, Chicago.

CHAPTER 9

ORTHOPEDIC IMPLANTS

Robert M. Rose

The surgical use of metallic fixtures to correct biological or mechanical accidents is at least 400 years old, as Petronius is known to have corrected the cleft palate in 1565 by the use of gold plates. The state of the art some two hundred years later is best described by a quotation from Venable and Stuck [1:3]:

The original dispute over the use of internal fixation of fractures was thoroughly aired in the French Surgical Journal of August, 1775. Monsieur Pujol, physician of Castres, took to task Monsieur Icart because the latter performed an open operation on a fractured humerus with disastrous results. Pujol in his best satire said, "It was about nine years ago that the nephew of a certain Seguier, mason of this village, had his arm broken by a cart and that they despaired of being able to save it for him, so great was the shattering. Then the surgeon, equally brave for conserving as for chopping off, presented himself to take charge of this treatment, and in order to hold in place the fractured pieces without the aid of any trying and cumbersome apparatus, he made deep sections longitudinally through the soft part and introduced immediately around the bone, in piercing the flesh, some brass wires of which he formed some several rings that were not at all delicate, and the ends of which he took care to twist together well. This beautiful maneuver had the result that one should expect, and gangrene, of which the patient died two days later."

M. Icart began his reply by stating that the operation took place fourteen years ago instead of nine years ago and "you (Pujol) were at Toulouse . . . at the time of the treatment of the mentioned Seguier . . . You would have then after an accident of this sort, gangrene succeed inflammation in two days time, and this gangrene kill the patient like a pistol shot . . . . I would have the honor meanwhile to tell you that I have seen brass wire put to use with success by the famous Lapeyode and Sicre, surgeons of Toulouse, whose lights and talents you have yourself respected. I myself would not be averse to believing that this wire of brass, silver, or gold would be applicable in certain cases, such for example as when in a fracture with the complication of

of large wounds, the bone is found denuded of flesh and of its periosteum ... Sometimes I say that in order to avoid accidents, it is
necessary to bring the pieces of the bone together and to hold them
either by means of a wire or a little band." At the end of the reply
Icart added a detailed case report of the pateint on whom brass wire
was used to mend the fracture and he stated that the patient died
twleve days after the accident. He presented the affidavits of two
surgeons to the effect that they had changed the dressing on the
damaged arm "eight hours before the death"and that the wound was in
excellent condition and "there was nothing in the patient' s state to
indicate approaching death. "

As inflammation, gangrene and the other medical factors attending

orthopaedic surgery were largely mastered, problems with the im-

plant materials themselves became apparent. Mechanically the im-

plants are relatively simple. For instance, Von Langenbeck and

Hansmann both had, about in the 1880' s, used metal plates to fix

fractures (cf. Figures 1-3).

Figure 1

Such "bone plates" are in common use today. They are used to fix

bone fractures in much the same way as broken stormdoors or sashes

are fixed in the home. The screws may be self-tapping, or the

surgeon may have to drill a hole in the bone (using brace and bit or

electric drill!) and then tap threads! Somewhat more sophisticated

are appliances such as the Jewell nail, which is used to fix fractures

of the hip, are shown in Figures 4-5.

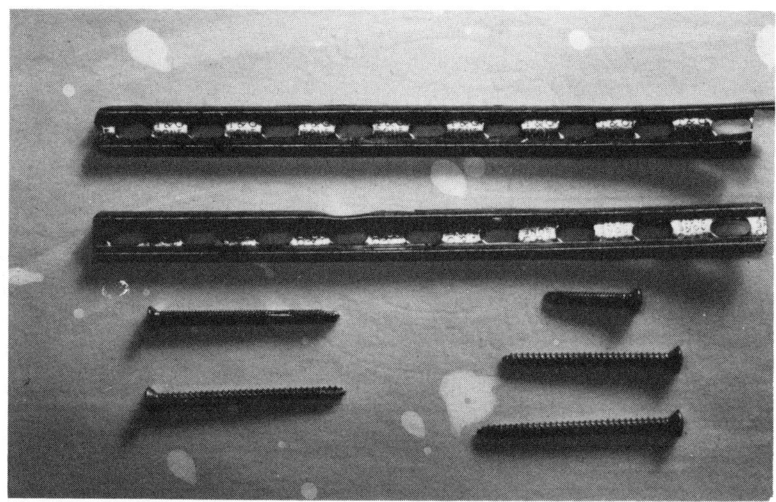

Figure 2

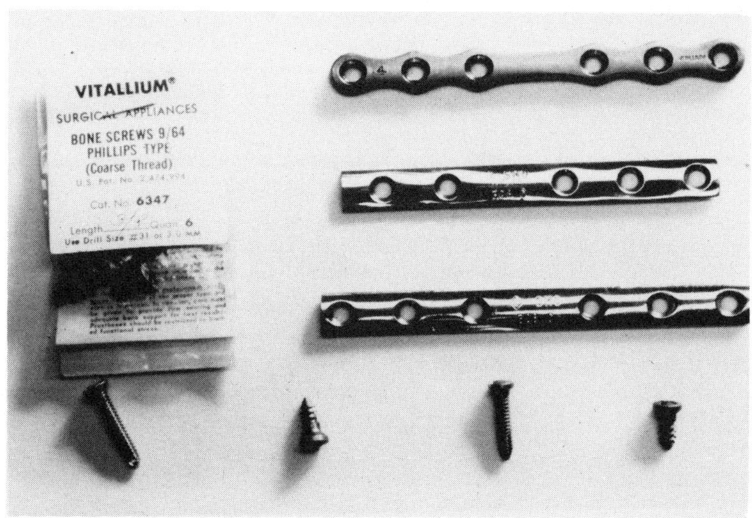

Figure 3

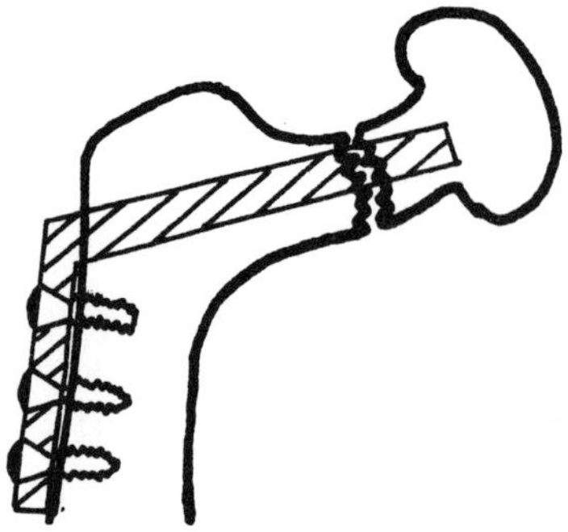

Figure 4

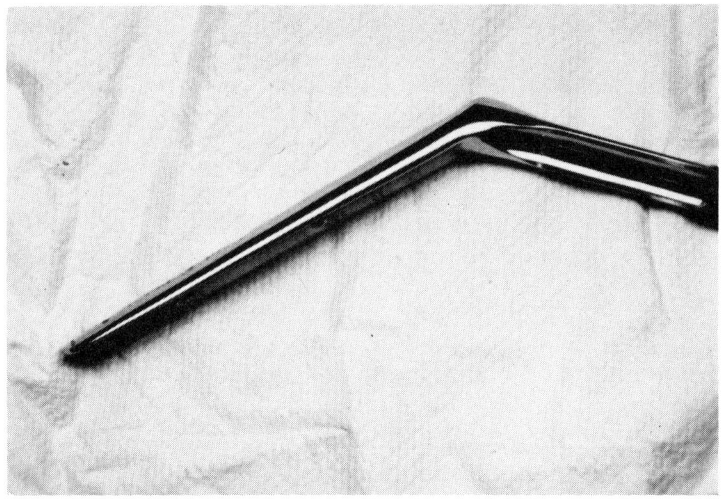

Figure 5

Sometimes, union of the fracture is dubious due to age or physical condition, and sometimes it is not desirable to keep the victim in bed very long; and sometimes the joint has been effectively destroyed by disease. In such cases the entire hip joint or the entire femoral head can be replaced (cf. Figures 6-8).

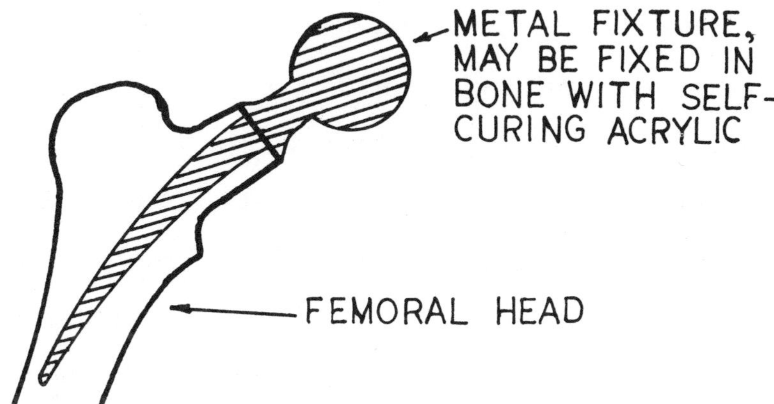

Figure 6

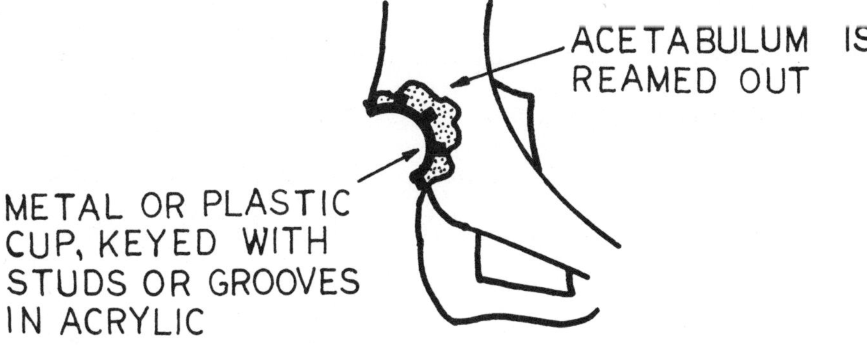

Figure 7

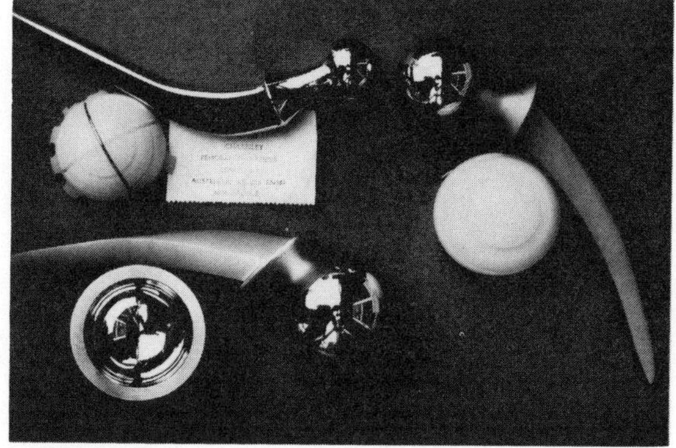

Figure 8

The femoral head is replaced by a metal arm which is installed in the shaft of the femur after the remains of the old head have been sawn off. The pelvic socket (acetabulum) can be reamed out and replaced by a metal or plastic cup. The photo shows three sets of hardware used to replace the hip joint, made of varying materials. The "total hip" prosthesis is only one example. Metals and plastics are used for many other purposes besides fracture fixation and skeletal prosthesis: for mesh for the restraint of hernias, for rebuilding jaws and anchoring dentures, to replace blood vessels and bile ducts, etc. Our attention today will be focused on the orthopaedic applications. Now there are a number of straight forward requirements we can see right away:

(a)   There has to be enough mechanical strength, not only static strength but fatigue resistance. Since bone has a tensile strength of about 10 - 15,000 psi, we look for about 100,000 psi in the implant material.

(b)   The resistance to corrosion must be extremely high.

(c)   Toxicity, carcinogenicity, and allergenic potential should be low or nonexistent.

These requirements, taken together, are very stiff. Early implants were made of plain carbon steel, aluminum, whatever was handy, even brass -- with messy results. Blood plasma and body fluids are saline (.1 - 1 molal NaCl) with dissolved CO_2, O_2, urea, all sorts of interesting organic acids as well as proteins. Great stuff for corrosion testing! And, corrode is what those implants did. In such cases, the corrosion products can kill the surrounding tissues, and pain,

inflammation, pus and worse results. Eventually, vanadium steel

was used, then stainless steels of various kinds, and then cobalt-

chromium alloys. Today, the main materials in use are vitallium

and HS-25, both basically Co-Cr alloys, and stainless steel of the

316L-317L type. The compositions and typical mechanical properties

are shown in Table 1. The two cobalt-chromium base alloys HS-25 and

vitallium are essentially "stellite" high-temperature alloys (HS:

Haynes Stellite) and in fact vitallium is a very close relative of HS-21.

In 1965 Cohen reviewed the principal causes of failure in 300 cases

[2:354], which, from the most to the least frequent were found to be:

1.   conception inadequate to function
2.   stress raisers (notches, holes et al.)
3.   inadequate material strength
4.   desired mechanical properties were not achieved by the fabrication methods used
5.   improper choice of material for adequate corrosion resistance
6.   improper fabrication method for proper corrosion resistance
7.   mechanical defects due to processing laxity
8.   poor corrosion resistance due to processing laxity
9.   contamination of metal or metal surface
10.   improper labeling (dimensions et al. !)

From the materialists' point of view, there is a study by Cahoon

and Paxton [3:1] which traced various failures back to:

(a)   inclusions, MnS and otherwise, in stainless steel, which started

fatigue cracks. Sometimes these inclusions were in sharp corners!

(b)   shrinkage porosity in vitallium

(c)   holes drilled very close to the edges of stainless bone plates; the

material between hole and edge was thus transformed from ductile

Table 1.   Compositions and Mechanical Properties of Common
Orthopedic Implant Materials.

material	stainless steel (316L)	HS-25	Vitallium
Cr	17 - 20%	19 - 21%	25 - 30%
W	----	14 - 16	----
Ni	10 - 14	9 - 11	*
Mo	2 - 4	----	5 - 7
C	.08 max.	.15 max.	.35 max.
Mn	2 max.	2 max.	1 max.
Si	.75 max.	1 max.	1 max.
Fe	base	3 max.	*
Co	----	base	base
P	.03 max.	.03 max.	----
S	.03 max.	.03 max.	----
B	----	----	.007 max.
Mg	2 max.	1 - 2	1 max.
YS(ann)	30 ksi	~ 50 ksi	80 ksi
YS(c.w., s.r)	120 ksi	up to 190 ksi	***
UTS (ann)	80 ksi	~ 130 ksi	100 ksi
UTS (c.w., s.r)	140**ksi	up to 250 ksi	***
R.A.	30 - 50%	10 - 50%	8 %

*   Ni + Fe less than 1%
**   up to 200 kpsi is possible by cold work.
***   not supplied in this condition — low ductility!

FCC austenite to brittle BCT martensite. In this case, as it turned out, improper heat treatment had "sensitized" the stainless steel and it had very poor corrosion resistance. Luckily, it broke before it could be implanted.

The metallurgy of stainless and stellites has improved to the point where there is very little excuse for inclusions or porosity, and the enemy that remains is corrosion.

It is easy enough to see the elementary corrosion cell (cf. Figure 9) involving the reactions

$$Fe \rightarrow Fe^{++} + 2e^{-} \qquad\qquad \xi = .44 \text{ Volts}$$

$$Cu \rightarrow Cu^{++} + 2e^{-} \qquad\qquad \xi = -.345 \text{ Volts}$$

gives us a net driving force of 0.785 volts, or, via

$$\Delta G = -n\xi F$$

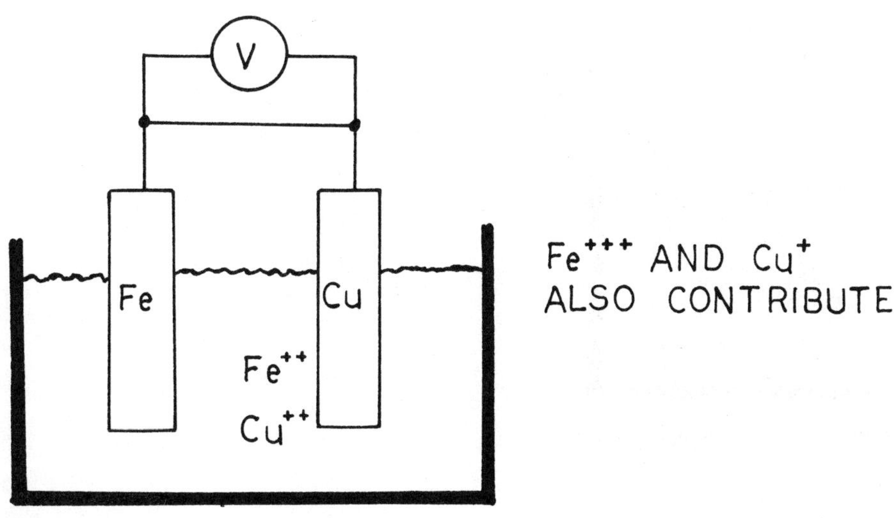

Figure 9

an appreciably negative free energy. Obviously any heterogeneity, strains, plastic deformation, composition, can give us a net ΔG between two points in a nominally homogeneous alloy; dissimilar metals are not needed. In such a case, say (cf. Figure 10)

$$Fe \text{ (strained)} \rightarrow Fe^{++} + 2e^{-}$$

is obviously the anode (oxidation) reaction, but the cathode (reduction) reaction doesn't have to be

$$Fe^{++} + 2e^{-} \rightarrow Fe \text{ (unstrained)}.$$

This is important, because the Fe^{++} may not make it to the cathode. Instead, in acid media,

$$H^{+} + e^{-} \rightarrow 1/2\ H_2$$

or in neutral or alkaline media

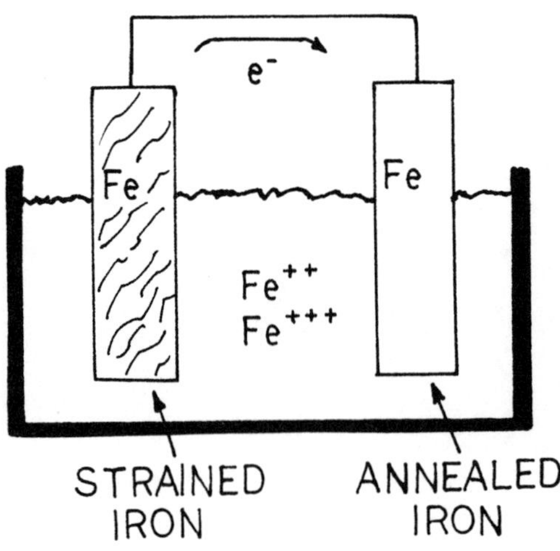

Figure 10

$$O_2 + 2H_2O + 4e^- \rightarrow 4OH^-$$

can occur at the cathode.

Now we have to consider mass action, since

$$\Delta G = \Delta G^o + RTnQ$$

implies concentration dependence. Thus, the presence of more Fe^{++} in solution tends to decrease the anode reaction voltage. In fact, we can recognize that a piece of perfectly homogeneous Fe will corrode if it is exposed to different concentrations of Fe^{++} at different points. We can also bring on corrosion by varying the cathode reaction voltage at two different points:

$$O_2 + 2H_2O + 4e^- \rightarrow 4OH^-,$$

$$\xi + \xi^o + \frac{RT}{nF} \ln \left[\frac{C^4(OH)}{Po_2} \right]$$

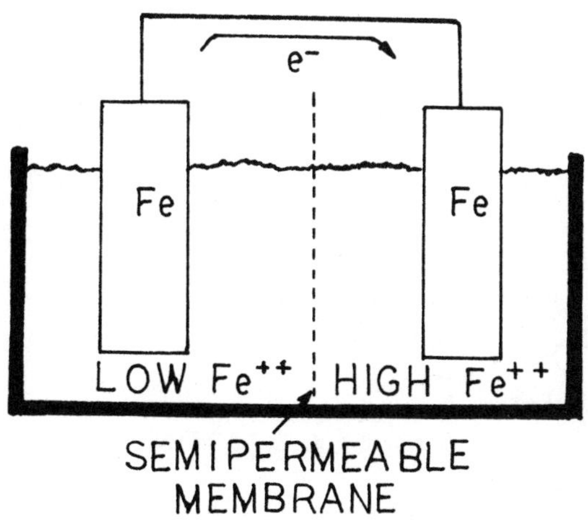

Figure 11

we can vary the OH^- concentration or the dissolved oxygen. The

presence of dissolved oxygen tends to push the cathode reaction along,

and if we have one place poor in oxygen and one place rich, say the

top and bottom of a bolt head, we should expect the oxygen-rich area

to be a good cathode, and the other place, the anode (Fig. 12). Corro-

sion does occur <u>under</u> the bolt-head, at the anode. As a general rule,

<u>corrosion occurs where the oxygen isn't.</u> Even in acid solution, this

rule applies, because the reduced hydrogen "plates out" on the cathode

and thereby blocks further reaction (this is called polarization).

Dissolved oxygen solves this problem, possibly in a number of ways.

As it turns out, polarization or mass action effects at the anode are

much less serious (you can put copper nails in a steel roof with little

harm, but steel nails in a copper roof are very short-lived). So,

"crevice corrosion" via oxygen concentration differentials is a buga-

boo. (Here, for instance, is corrosion which occurred under barn-

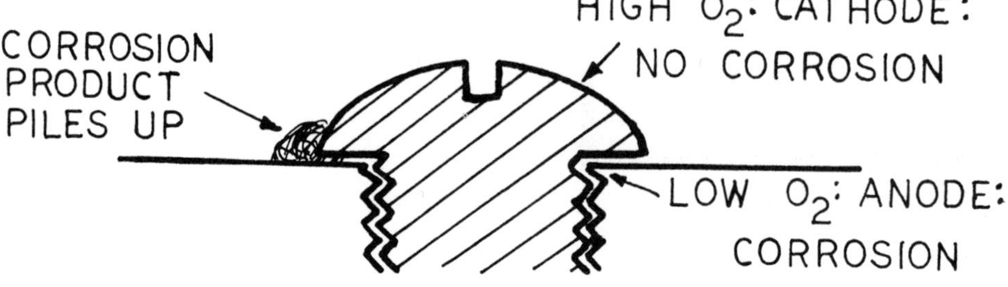

Figure 12

Figure 13

acles.) From Figure 13,   cross sections of screws in bone plates, we

see the possibilities are apparent: there is enough variation that there

is no way to predict the dimensions of the crevice.   There are also

great variations in the holding power of the screw, which can loosen

or stay put due to changes in the underlying bone.

As it turns out, the crevice corrosion story is more complication

for "passive-film" alloys such as stainless steel and stellites, which owe their excellent corrosion resistance to tightly adherent surface layers of oxygen. In such cases, the <u>local</u> concentrations of H^+ and Cl^- shoot up sharply, the protective film in the crevice is destroyed, and corrosion pits form. Stainless steel is particularly susceptible to this mechanism, and in 1969 Colangelo and Greene examined 53 devices made of 316L stainless which had been removed for a variety of reasons. In 24 there were corrosion. More important, of the 53 devices, 23 were multicomponent (screws and plate, etc.), and of the 23, all but two were corroded [4:247].

In view of the crevice-corrosion problem, the question arises as to why surgeons bother with 316L stainless at all. Part of the answer lies in the ductility of stainless and the brittleness of vitallium. The surgeon can bend a stainless plate if he has to. Also vitallium must be cast and therefore the range of shapes is limited and there is always the casting porosity problem. (I suspect that there are cases where bone actually grows into such porosity, making implant removal a very difficult proposition.) So, although vitallium has much better crevice corrosion resistance than 316L stainless steel, the latter is still wide-ly used. Of course, there is the third alloy on our table, HS-25, which is ductile. It is also sold under the trade name "vitallium", and is in fact more resistant to crevice corrosion than 316L stainless, but considerably less so than vitallium. Sometimes HS-25 and vitallium are combined in an implant; both components are stamped "vitallium" and the surgeon is no wiser. It has been a matter of conjecture whether dissimilar metal effects are important in such a case, and galvanic currents measured under ideal conditions are negligible, for combinations of HS-25, vitallium and titanium. However, such experiments have not been made with crevices, or with the stress concentrations you can't avoid sometimes. Recently we found evidence [5] that these factors together with dissimilar metals are troublesome. Figure 14 is an x-ray of a broken nail-plate assembly which was implanted in an attempt to fix a broken hip (cf. Figure 14).

We found that the nail was made of vitallium and the plate was HS-25. However, the plate had literally snapped off, despite the considerable ductility of HS-25 under ordinary conditions. After some

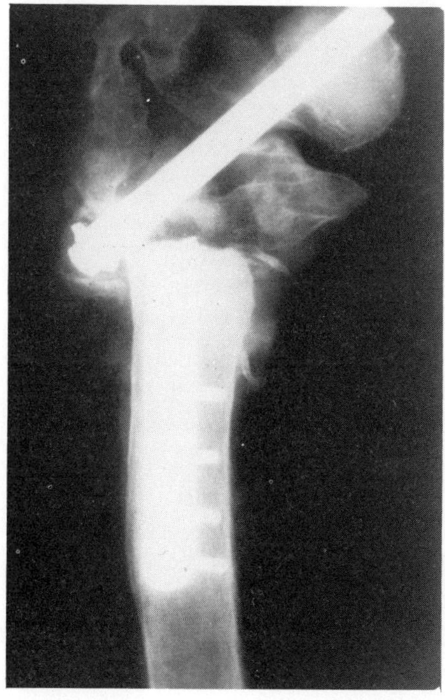

Figure 14

searching, we identified the culprit as corrosion fatigue. Figures 15-

17 show   one of the many corrosion pits which nucleated the fatigue

cracks, one of which then skipped gaily across the specimen in a

series of successive cleavages.

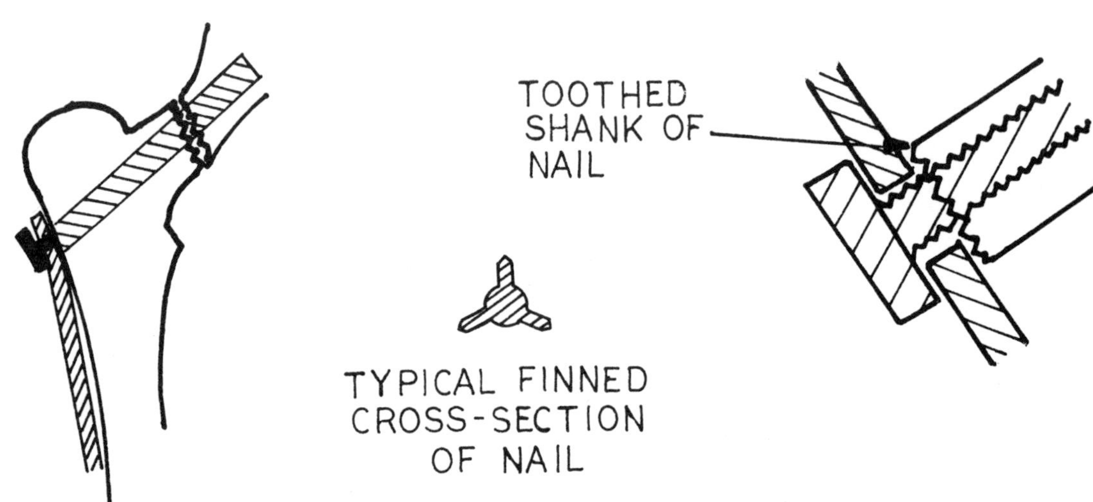

Figure 15

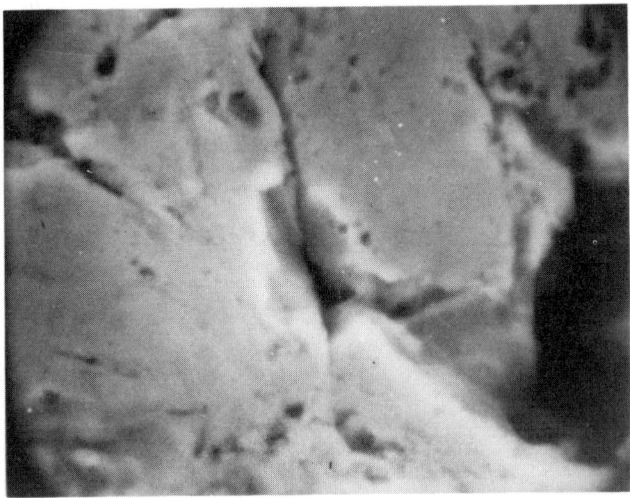

Figure 16

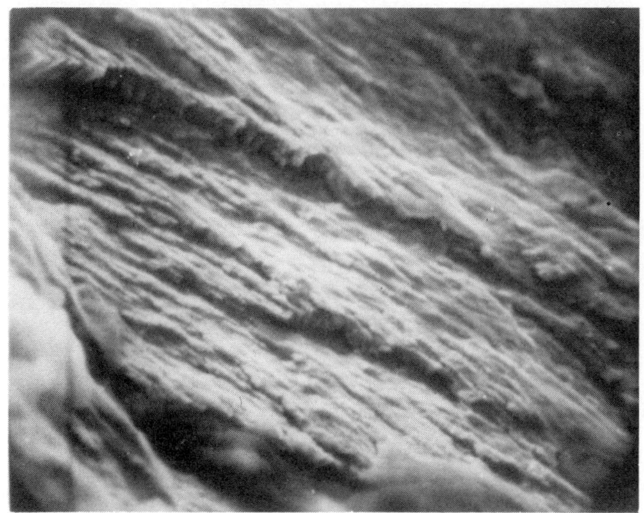

Figure 17

There is also some prior evidence [6] that HS-25 does indeed pit

corrode in the body, and unpublished work by Wulff indicates that HS-

25 lies far below vitallium in the hierarchy of alloys resistant to crev-

ice corrosion.

There are alternate materials, but not nearly as popular or well-

accepted as the three we have been discussing. There are titanium

alloys, particularly Ti-6Al-4V which is very resistant to corrosion

and can be heat treated to a 160,000 psi yield stress. Tantalum can

also be used if great ductility is needed.

We should observe several qualifications to the corrosion problem:

(a) Sometimes the body may successfully form pseudo-membranes

and isolate even very irritating metallic intrusions. Thus war veter-

ans may walk about with many shrapnel pieces inside them and with no great discomfort. Even with vitallium; cobalt, chromium and

molybdenum do tend to leach out into the surrounding tissues, and

Ferguson et al. [7:77] find about 150 ppm total of these elements

there.

(b)   Sometimes corrosion may be desirable. For instance, corrosion

of silver amalgam seals the margin of a tooth filling. Without this

corrosion, decay would undermine the filling.

Another set of problems attend the total hip prosthesis. For the all-

metal prosthesis, the ball, although finely polished, (Fig. 18), has a

relatively high friction coefficient, which leads to great mechanical

difficulties. The friction force for a 150 lb. man can in fact be more

than 150 lb. , due to the geometry of the hip. The prosthesis may

loosen at either end. The best way to anchor both parts in femur and

pelvis is to use blobs of self-curing polymethyl methacrylate, mixed in

Figure 18

the operating room [8].   The cup has prongs to anchor it in the meth-

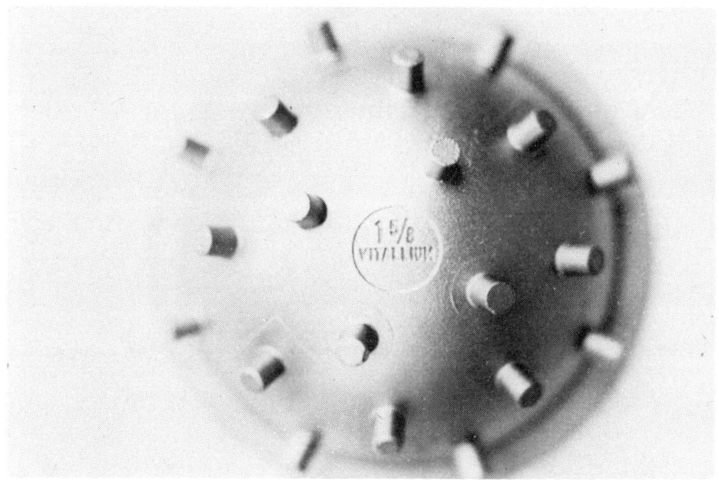

Figure 19

Figure 20

acrylate. The methacrylate has the advantage of distributing the load on the prosthesis over a wide area, avoiding recession problems with the bone; see Figures 19-20. (It also kills the adjacent bone, which retards recession considerably!) However, the acrylic monomer may in fact be carcinogenic.

As a solution to the friction problem, Charnley has used a small stainless ball and a high-density polyethylene cup (this device is shown in Figure 8) but the ball literally wears through the cup in many cases. (The other plastic-metal prosthesis shown in the fourth photo is a Charnley-Mueller total hip, which has a somewhat larger ball and uses vitallium rather than stainless for the femoral head.)

At present there are no bright promises in this area. Teflon has thus far proven to be a failure as an alternate cup material, and we can add friction and wear to corrosion as a prime problem. Let me end with a poem by G. Maurice Down (I think) entitled "To Titanium",

O lustrous metal worth thy weight in gold
Close to my femur's cortex to reside,
Restore its contours, shattered fragments hold;
Within my periosteum preside
Inert. Whilst there no toxic taint impose
On tissues torn in trauma's twisting track,
Lest cased in P.O.P. from front to back
I listless lie in search of sweet repose.
In day-dream softly I thy mistress call,
Imploring all Titania's fairy aid
That screws nor plate by chance be disarrayed;
So hope that I once more around may crawl.
Ambulant proof that every little helps –
By courtesy of Down Bros. Mayer and Phelps. [9]

REFERENCES

[1]   Venable, C. S. and W. G. Stuck.   The Internal Fixation of
Fractures. Springfield, Illinois: Thomas, 1947.

[2]   Cohen, J. J. Mater. 1:354.

[3]   Cahoon, J. R. and H. W. Paxton. 1968. J. Biomed. Mater.
Res. 2:1.

[4]   Colangelo, V. J. and N. D. Greene. 1969. J. Biomed. Mater.
Res. 3:247.

[5]   Rose, R. M. , A. Schiller, and E. Radin. 1971. To be published
in J. Bone and Joint Surgery.

[6]   Cohen, J. and J. Wulff. 1971. To be published in J. Bone and
Joint Surgery.

[7]   Ferguson, A. B. , P. G. Laing and E. S. Hodge. 1960. J. Bone
and Joint Surgery. 42A:77.

[8]   Charnley, J. 1970. Acrylic Cement in Orthopaedic Surgery.
E. and S. Livingstone.

[9]   Down, G. M. 1965. Lecture to Inst. Brit. Surgical Technicians,
Caxton Hall, 23 November 1965.

CHAPTER 10

BLOOD PRESERVATION BY FREEZING

Charles E. Huggins

Blood transfusion is now a widely practiced and generally safe form of
medical therapy. The science of blood group immunology has advanced
over the past seventy years to a point at which serious or fatal reactions are now a rarity. Advances in preservation techniques permit
blood to be collected and stored in the refrigerator for twenty-one
days.

Four major problem areas remain:

1. Supply and Demand: The traditional approach to the clinical problem of inventory control has been based on Random Collection for
Random Need. Although sound in theory, the twenty-one day dating
period is too short to fully equilibrate random and seasonal fluctuations
in availability and need for blood. Imbalance is always greatest with
the "low frequency" blood groups, e. g. , Group B (10 percent), Group
AB (5 percent), and all Rh-negative groups (15 percent) of the random
donor population. Critical shortages of blood during the summer and
at New Year's alternate with excess during the spring and fall. Nationally, it has been estimated that twenty percent of the six to seven
million units of blood collected annually are lost by outdating.

2. Metabolic: Since blood can be kept for twenty-one days, the belief
has arisen that the cells are equally effective at all times between

collection and outdating. Deterioration of the cells during storage in the regrigerator is now recognized clearly, and undesirable clinical side-effects following massive transfusion with aged, but "indate" blood are apparent.

3.   Immunologic: Although red cell typing and compatibility testing are highly developed, "Reactions" to specific white blood cell antigens and plasma protein components are now recognized commonly, and may be very serious.

4.   Infectious: Although syphilis, malaria, and other infectious diseases can be transmitted by blood transfusion, SERUM HEPATITIS is a public health problem of first magnitude. Hepatitis is said to follow 0.1 to 1.0 percent of all transfusions; and 10 percent of the clinical cases are fatal. Nationally, there are estimated to be 30,000 cases of serum hepatitis annually. Each patient requires an average of thirty-five days additional hospitalization, which at $100/day represents a yearly cost of $100,000,000 to treat hepatitis.

Solution to these problems of blood transfusion lies in two areas: PRESERVATION PLUS PURIFICATION - preservation to arrest the aging of blood cells, and purification to remove unnecessary, unwanted, and potentially injurious elements from the donor blood.

PRESERVATION AND FREEZING INJURY

As late as 1683, Robert Boyle wrote, "Since Coldness being a Tactile Quality, it seems impertinent to seek for any judges of It than the

Organs of that sense, whose proper object it is". The concept of cold as a subjective sensation contrasts remarkably with the highly developed knowledge of astronomy, and other physical sciences, at that time.

Fahrenheit's invention of the thermometer (1709), Celsius' temperature scale (1742), Maxwell's equations for calculating the average velocity of molecules in an ideal gas (1860), and Kelvin's concepts of absolute zero were remarkable advances in physical science since the time of Robert Boyle.

Advances in biological sciences, however, did not follow in parallel. Numerous investigators tried to freeze isolated cells and whole animals. They were able to confirm, with dismal regularity, the finding that freezing and thawing of living mammalian tissues was an invariably lethal event.

Since ice formation in tissue was followed by death on thawing, and since ice crystals have sharp points, it was assumed (quite reasonably) that the ice crystals pierced, or otherwise physically damaged, the delicate cells. This concept was strengthened by the observation that extremely rapid freezing and thawing of water produced an acrystalline or "vitreous" form of ice. Under similar conditions a significant percentage of human red cells escaped injury.

Physical chemists have long recognized that when ice forms in a salt-containing solution the crystals are pure water. Dissolved solutes become concentrated until the eutectic temperature is reached, at which point no further concentration takes place.

In 1929 Moran suggested, and in 1953 Lovelock demonstrated [1:414], that the concentrated solute formed during freezing in a biological system can be highly toxic. Numerous investigators have now confirmed that human red blood cells become damaged when suspended in sodium chloride solution stronger than 0.8 M. NaCl (Figure 1). It is possible to reproduce many of the effects of freezing merely by suspending cells or tissue in a hypertonic environment [2:483].

Although excessive concentrations of solutes by themselves can damage living cells, many other physical-chemical systems are also

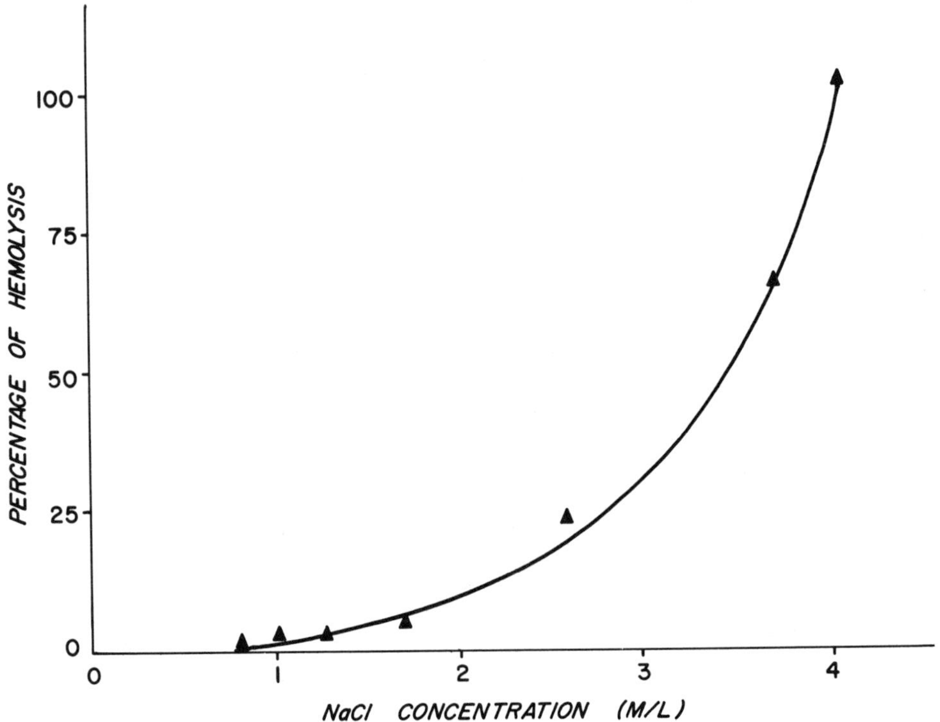

Figure 1

affected by the solute concentrating action of freezing [3:427]. Crystallization of one member of a two-salt buffer system, for example,
alters the relative concentrations of the two salts and thereby changes
the pH. Protein solutions are also important since freezing shifts
their pH toward the isoelectric point. The final pH of a frozen biological system may be widely variable and is the result of changes in
different directions in many different systems.

Lipoprotein may be denatured by freezing with removal of structur-
ed water that links the fat and protein components. Denaturation of
lipovitellin of egg yolk, for example, occurs during freezing in an
aqueous medium without salt [3:427].

With human erythrocytes, destruction of membrane lipoprotein can
be demonstrated either by freezing or by suspending the cells in concentrated salt. Destruction of membrane lipoprotein may be the
proximate cause of hemolysis during slow ($\Delta T < -1^{\circ}C$ per min or less)
freezing and thawing.

Modification of the structure of ice and changes in the concentration
of solutes during thawing make freezing and thawing inseparable. One
cannot be considered without the other. Hopefully, the refined techniques of cryomicroscopy of Diller and Cravahlo will help to clarify
these problems [4:191].

PRESERVATION - CRYOPHYLAXIS

The most significant advance in practical aspects of low temperature biology since Robert Boyle came with the chance discovery that glycerol had the ability to protect living cells against freezing injury [5:666]. Glycerol freezing techniques have now been applied with varying degrees of success to preservation of isolated cells, small bits of tissue, whole organs, and animals.

Chemical analogues of glycerol have been found to share the ability to protect against freezing injury (Figure 2). Of these, some protect to all temperatures, others - e.g., methanol and ethanol - do not.

A class of chemical compounds -- Endocellular Cryophylactic Agents (ECA) -- has now been recognized, each member of which has the following general properties:

1. small molecule
2. miscible with water in all concentrations
3. poor fat solvent
4. very high boiling point

Those compounds with boiling points below that of water <u>do not</u> protect to all temperatures and are weak ECA's. Those with boiling points significantly above 100°C can protect to all temperatures, and are strong ECA's. Dimethylsulfoxide (DMSO) was the first chemical compound whose cryophylactic properties were predicted from its known physical and chemical properties [6:1394].

ECA's apparently act by their ability to bind water molecules, thereby interfering with the orderly formation of ice. At any temperature the percentage of solvent water sequestered in ice is reduced, consequently reducing the degree to which dissolved solutes concentrate [2:483].

STRONG ECA'S   WEAK ECA'S

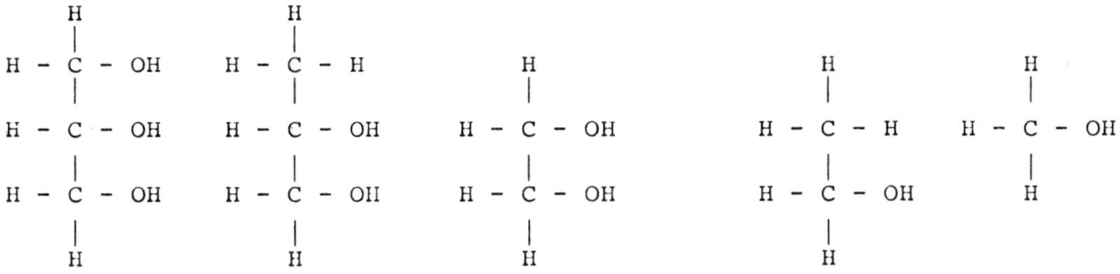

```
    H                H
    |                |
H - C - OH       H - C - H           H                   H                H
    |                |               |                   |                |
H - C - OH       H - C - OH      H - C - OH          H - C - H        H - C - OH
    |                |               |                   |                |
H - C - OH       H - C - OH      H - C - OH          H - C - OH           H
    |                |               |                   |
    H                H               H                   H
```

Glycerol   Propylene   Ethylene   Ethanol   Methanol
                  glycol   glycol

BP=290°C   BP=189°C   BP=197°C   BP=79°C   BP=65°C

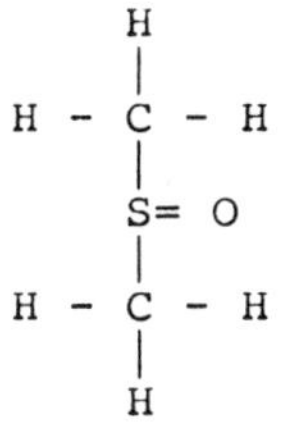

```
    H
    |
H - C - H
    |
    S= O
    |
H - C - H
    |
    H
```

Dimethyl
sulfoxide

BP=189°C

Figure 2

Action of these compounds is based on their ability to form coordinate bonds with water, rather than on specific chemical action as illustrated by results of the following experiment: [7:5190]

Equal volumes of 8.6M ECA - Saline, and human red cells were mixed, frozen to $-85^{\circ}C$, stored for 24 hours, and thawed.

Expt	ECA	Hemolysis on thawing
1.)	8.6M Glycerol	< 5 percent
2.)	8.6M Propylene glycol	< 5 percent
3.)	8.6M Ethylene glycol	< 5 percent
4.)	8.6M DMSO	< 5 percent
5.)	2.15M Glycerol 2.15M Propylene glycol 2.15M Ethylene glycol 2.15M DMSO 8.60M ECA	< 5 percent
6.)	Water	100 percent

To be effective, ECA's must cross the cell membrane with ease. Under steady state conditions, therefore, they exert no net osmotic effect [8:133]. Solute must be incorporated in ECA solutions to maintain normal cell size. In fact, cell size in ECA solutions is dependent entirely on the concentrations of other solutes across the cell membrane.

If the solute is hypertonic -- small cell size -- hemolysis on thawing is excessive. If the solute is hypotonic -- large cell size -- progressive hemolysis occurs during storage. ECA's in concentrations greater than 6.0M are toxic. When concentrated solutions are used, the ECA must be added gradually to the cells to effect a smooth rise in concentration. The opposite order of mixing is likely to be very damaging

[2:483].

Transfer of ECA's across the cell membrane occurs both by passive and by facilitated diffusion. The latter transport mechanism requires active cellular metabolism, and is blocked by cold. In general, temperatures between 15° and 25^oC provide the optimum balance between ECA transport, and _ex vivo_ cellular deterioration.

The rates of freezing and thawing can act synergistically with the ECA concentration. Under extremely carefully controlled conditions of rapid freezing and thawing (approximately $100^oC/sec.$) it is possible to obtain significant recovery of human red blood cells without ECA's.

Slower rates of freezing and thawing ($0.1 - 1.0^oC/sec$) can be used satisfactorily if ECA's are present in 1.5 to 3.0M concentration.

Very slow rates (0.1 to $10^oC/min$) of freezing and thawing are tolerated satisfactorily if ECA's are present in 3.0 to 5.5M concentration.

In general the most critical temperature range is between -1° and -40^oC. Experience has shown that the higher the final concentration of the ECA the greater the variation in storage temperature that can be tolerated with safety.

Thawed ECA-containing-blood cells, if transfused directly, hemolyze, or burst. Rupture of the cell membrane is caused by rapid endosmosis of water that cannot be balanced by equally rapid transfer of ECA.

Gradual removal of the cryophylactic agent after thawing must be done to balance osmotic relationships across the cell membrane. Destruction of blood cells during ECA removal is a non-specific osmotic effect. Hemolysis at this stage, however, is as undesirable as hemolysis during freezing and thawing.

Glycerol and other ECA's form very firm hydrogen or coordinate bonds to water. This bio-physical property is the principal reason for their effectiveness in preventing cellular death during freezing and thawing. A corollary of this action, however, is the necessity to re-move these compounds by dilution in a large aqueous volume. Techni-ques that have proved effective for glycerol removal all act on the principle of dilution of the ECA complexed to water. Separation meth-ods designed to separate ECA's from water (partition, precipitation, or fractional distillation) have proved ineffective.

Gradual dilution and dialysis were the first methods used for re-moval of glycerol from thawed blood. Both methods are effective, but neither removes free hemoglobin or red blood cell ghosts. Serial washing by centrifugal techniques is effective, but time consuming and laborious: free hemoglobin and ghosts are removed, but the procedure is exacting and each wash brings the danger of breaking the blood container or contaminating the cells.

Special centrifuge heads were designed by Brown and Hardin and by Mollison et al. in their attempts to simplify and automate serial wash-ing techniques [9:267] [10:256]. Unpredictably, uneven flow rates of the wash solutions through the cells led to disuse of these ingenious devices. Continuous flow centrifugal washing has been performed by the Arthur D. Little Company Cohn fractionator. In operation the erythrocytes are layered around the periphery of a rapidly spinning conical centrifuge bowl. Wash solutions pumped through a centrally located inlet port pass through the layered cells and exit by way of another port. Although excellent preservation of erythrocytes was obtained the method has not been used widely [11:161] [12:773].

Attempts have been made to compromise between rapid freezing and slow freezing glycerol techniques in the hope that reduced concentrations of ECA would reduce the problem of washing. These techniques require <u>both</u> control of rates of freezing and thawing, and post-thaw washing. Cells preserved by these techniques have been shown to have excellent survival [13:184] [14:176].

We studied the phenomenon of reversible agglomeration of erythrocytes in aqueous nonelectrolyte solutions in a search for new methods for removal of glycerol and other ECA's from thawed blood [15:504]. One volume of packed human blood was suspended in 10 volumes of isotonic sucrose. Within three minutes, progressively larger and larger clumps of cells formed and these settled rapidly to the bottom of the container. Equally as amazing as the observation of clumping was the further discovery that resuspension of the cells occurred after decantation of the supernatant and the addition of an equal volume of saline, or plasma. The phenomenon has been called reversible agglomeration to distinguish it from agglutination and aggregation, both of which have acquired specific hematological connotations in terms of mechanism.

Agglomeration appears to be a two stage process in which gamma globulins form a complex with lipoprotein in the wall of red blood cells at a pH optimum between 5.2 and 6.1. In the second stage as the ionic strength is lowered, gamma globulin precipitation causes co-precipitation of the red blood cells. Resuspension of the agglomerated cells takes place either by raising the ionic strength, or by raising the pH.

The speed with which ECA treated thawed blood cells may be diluted depends greatly on the rate of transfer of the ECA across the cell membrane. Dimethylsulfoxide and ethylene glycol move quickly, and can be removed readily. Unfortunately, these agents cannot be used for clinical blood preservation because of possible toxicity _in vivo_.

Glycerol, on the other hand, transfers relatively slowly across the cell membrane, but can be used for clinical blood preservation because it is a non-toxic normal body constituent.

When glycerol is used to preserve blood by freezing, the initial step in deglycerolization usually consists of crenation by addition of strongly hypertonic solute (50 percent Dextrose or 12 percent chloride). If done appropriately the cells are shrunken transiently, thereby reducing the absolute amount of intra-cellular glycerol, and increasing the margin of safety during subsequent osmotic swelling.

Protection of living cells during freezing, storage, and thawing is largely a function of the ECA concentration, whereas ease of removal of an ECA from cells after thawing is largely a function of the absolute amount of ECA inside the cell.

CLINICAL BLOOD PRESERVATION

The foregoing principles have been used to preserve human blood for transfusion by freezing.

Experience in many laboratories has shown that Red Blood Cells (Human) Frozen, the official name for glycerolized blood cells in the frozen state, may be kept for five to ten years. The low temperatures arrest metabolic deterioration during storage. This capability has been used for storage of exceedingly rare types of blood cells, and more recently in our laboratory for day to day regulation of inventory [16:138].

Experience with Red Blood Cells (Human) Deglycerolized, the official name for the thawed, deglycerolized blood cells, has shown good survival and function on transfusion. Oxygen carrying ability (the raison d'etre of blood cell transfusion) is equivalent to that expected from the time the blood was kept between 1° and $6^{\circ}C$ before freezing.

Leukocytes from the original donor are removed to a great extent during post-thaw deglycerolization. This purification of the red blood cells is reflected by significant reduction in both the incidence and severity of anti-leukocyte reactions in previously sensitized recipients.

Plasma removal during deglycerolization permits group O Red Blood

Cells (Human) Deglycerolized to be used electively as "Universal

Donor" for recipients of any ABO group.   In addition, the incidence

of post-transfusion hepatitis appears to be very significantly reduced

in relation to that expected with whole blood, or packed blood cells.

It is likely that freezing techniques will play an increasingly import-

ant role in clinical blood transfusion practice.

REFERENCES

[1]   Lovelock, J. E.   1953.   The haemolysis of human red blood cells
by freezing and thawing.   Biochem. Biophys. Acta 10:414.

[2]   Huggins, C. E.   1963.   Prevention of hemolysis of large volumes
of red blood cells slowly forzen and thawed in the presence of dimethylsulfoxide.   Transfusion 3:483.

[3]   Lovelock, J. E.   1957.   The denaturation of lipid-protein complexes as a cause of damage by freezing.   Proc. Roy. Soc. [Biol.]
147:427.

[4]   Diller, K. R. and E. G. Cravalho.   1970.   A cryomicroscope for
the study of freezing and thawing processes in biological cells.   Cryobiology 7:191-199.

[5]   Polge, C. , A. U. Smith, and A. S. Parkes.   1949.   Revival of
spermatozoa after vitrification and dehydration at low temperatures.
Nature (London) 164:666.

[6]   Lovelock, J. E. and M. W. H. Bishop.   1959.   Prevention of
freezing damage to living cells by dimethylsulfoxide.   Nature (London)
183:1394.

[7]   Huggins, C. E. 1965.   Preservation of organized tissues by
freezing.   Fed. Proc. 24:5190.

[8]   Huggins, C. E.   1966.   Frozen blood: principles of practical
preservation.   Mono. Surgical Sci. 3:133-173.

[9]   Brown, I. W. , Jr. , and H. F. Hardin.   1953.   Recovery and in-
vivo survival of human red cells.   Studies of red cells after storage up

to six and one-fourth months at sub-zero temperatures. <u>A. M. A.</u>
<u>Arch. Surg.</u> 66:267.

[10]   Mollison, P. L. , W. C. Lister, A. F. Roth, and F. Smith.
1958.   Centrifugal washing of red blood cells. <u>Brit. J. Haemat.</u>
4:256.

[11]   Tullis, J. L. , J. G. Gibson II, M. T. Sproul, R. J. Tinch, and
P. Baudanza. 1970. Advantages of the high glycerol mechanical
systems for red cell preservation; a 10-year study of stability and
yield. <u>Modern Problems of Blood Preservation.</u> Spielmann, W. and
Seidl, S. Eds. Stuttgart: S. Gustav Fischer Verlag, 161-167.

[12]   Meryman, H. and M. Hornblower. 1969. A red cell freezing
and washing procedure using a high glycerol concentration. Abstract
in <u>XII International Congress on Blood Transfusion Abstracts.</u> Moscow:
MIR Publishers, 773-774.

[13]   Rowe, A. W. , J. B. Derrick, W. Miles, F. H. Allen, Jr. , and
A. Kellner. 1970. The biochemistry and clinical use of red cells
frozen by the low glycerol-rapid freeze technique. <u>Modern Problems</u>
<u>of Blood Preservation.</u> 184-198.

[14]   Krignen, H. W. , A. C. J. Kuivenhoven, and J. J. Fr. M. DeWit.
1970.   The preservation of blood cells in the frozen state. <u>Modern</u>
<u>Problems of Blood Preservation.</u>   176-183.

[15]   Huggins, C. E. 1963.   Reversible agglomeration used to remove
dimethylsulfoxide from large volumes of frozen blood. <u>Science</u> 139:
504.

[16]   Huggins, C. E. 1970.   Reversible agglomeration - a practical
method for removal of glycerol from frozen blood. <u>Modern Problems</u>
<u>of Blood Preservation.</u>   138-155.

CHAPTER 11

HEAT TRANSFER IN BIOMATERIALS

Ernest G. Cravalho

Man has been concerned with heat transfer in biomaterials since he first walked the face of the earth. His concern in these early days manifested itself in his quest for shelter which would afford him protection from adverse thermal environments and in the significance he accorded fire which would produce an "artificial" thermal environment that he could control to some extent. It is not surprising, then, that early civilization progressed most rapidly in those regions which provided a thermal environment most compatible with the thermal requirements of the human body. This compatibility enabled the members of these early civilizations to live and work efficiently in reasonable comfort and to minimize the amount of time and effort expended in providing the necessary thermal environment. Thus additional time was available to develop those aspects of society and culture that we presently use as indicators of progress in early civilizations.

In the milleniums that have ensued since these early civilizations, technology has developed to the point that man has extended his concern for heat transfer in biomaterials into domains more esoteric than his environment. To be sure, man is still concerned with the way his body thermally interacts with the environment but apart from his environment, man's concerns for heat transfer in biomaterials can be grouped into two very broad areas: (1) internal, in which he attempts to develop an understanding of the manner in which heat is transferred within the human body from both the physical and biological points of view and (2) external, in which he attempts to develop both the techniques and the hardware to control, to induce, or to observe by external means certain thermal states in the human body.

Research and development in these areas has been largely an interdisciplinary effort. To be sure, the earliest work was conducted primarily by biologists and physicians, but more recently engineers and physical scientists have been playing a more significant role. In many cases the really meaningful advances have resulted from collaboration between individuals from both the biomedical and physical sciences.

In the present treatment, we shall discuss briefly theory and applications of several aspects of heat transfer in biomaterials, dwelling at length upon certain selected topics.

APPLICATIONS

The applications of heat transfer in biomaterials are many and varied.

They include, but are certainly not limited to, the following:

1.   Heat transfer in biomaterials can be used to increase our basic
understanding of the physiological behavior of the human organism,
particularly as related to metabolism, circulation, and regulation of
body temperature. Such information can lead to an improved understanding of those thermal aspects of human physiology that can be
exploited for the purposes of diagnosis and treatment of disease.
Some of these aspects are listed below.

2.   Heat transfer in biomaterials is important in establishing the
nature of the interactions between the human organism and the environment. Included in these studies are the effects of temperature,
humidity, and metabolism on the physical comfort and well-being of
the organism.

3.   We are interested not only in the heat transfer phenomena that
affect the state of the whole organism but in those heat transfer
phenomena that occur in component systems as well. In these
studies we attempt to establish the effects of circulation, metabolism,
and tissue thermal properties on the heat transfer capability of
various components. A portion of this effort is directed at modeling
these heat transfer phenomena so that the thermal responses of living
tissue can be predicted analytically.

4.   In order to utilize the aforementioned models in a meaningful way,
it is necessary that accurate values for the thermal properties of
living tissue be available. To this end, considerable effort in the
study of heat transfer in biomaterials is expended in making measurements of properties such as thermal conductivity, thermal diffusivity,
and thermal inertia. This problem is not a trivial one but one re-
quiring special care to discriminate the effects of metabolism, circulation, geometry, and anisotropy.

5.   Considerable attention has been given in recent years to the preservation of tissues and organs by freezing. Heat transfer in these
biomaterials certainly plays a major role in the successful imple-

mentation of this technique. Important heat transfer processes occur
both within each of the individual cells that make up the tissue or
organ and within the structure of the tissue or organ itself. Research
is being conducted to distinguish between these phenomena and their
affects upon viability.

6. Sub-freezing temperatures can also be used to produce localized
necrosis in tissue, thereby selectively destroying undesirable cell
growths. This technique, known as cryosurgery, clearly benefits
from the findings of those who are trying to preserve tissue. The
major effort in cryosurgery is to establish the interdependence of the
various physical parameters and the manner in which these parameters limit the utilization of the technique.

7. Biomaterials, like any materials which have finite thermodynamic
temperatures, emit electromagnetic radiation. As we would expect,
the intensity and spectral character of the emitted radiation depends
upon the temperature of the material. In the case of a biomaterial,
this temperature is indicative of the physiological state of the biomaterial. Thus, by observing the intensity and spectral character of
the infrared radiation emitted by a biomaterial, we can infer a great
deal about the physiological state of the material. This technique,
known as thermography, can be used to diagnose metabolic anomalies
such as tumors and arthritis and circulatory anomalies such as block-
ages of arteries. Research in this area is devoted to the development of hardware to form thermograms rapidly and accurately and
to the characterization of metabolic and circulatory anomalies that
can be diagnosed by this technique.

8. Another technique for diagnosis which is limited by the ability of
biomaterials to transfer heat is ultrasound. High frequency sound
waves can be used to explore biomaterials for various anomalies;
however, a portion of the energy carried by these sound waves is
dissipated within the medium, and this dissipated energy can lead to
temperature excursions which can destroy biomaterials. The limitations that these temperature excursions impose upon dosage levels
depend in part upon the heat transfer characteristics of the medium.
Obviously, a material of high thermal diffusivity is capable of distributing the effects of a given dose of ultrasound over a larger
volume in less time than a material of low thermal diffusivity provided the energy absorption characteristics of the two materials were
identical. Obviously, ultrasound could also be used to produce local-

ized necrosis simply by exceeding the safe dose limit. Research on this subject is being devoted to the characterization of the interaction between the ultrasound and the biomaterial and the manner in which the thermal characteristics of the biomaterial influence the utilization of the technique.

9.   There is a class of thermal interactions between biomaterials and the environment which may be termed extracorporeal. Typically, these situations involve the displacement of a biomaterial from its natural surroundings to "unnatural" surroundings for some specific purpose. While in its artificial surroundings, it may be necessary to provide the proper thermal environment in order for the bio-material to survive. For example, in certain surgical procedures it is necessary to circulate the blood outside the human body. While in this extracorporeal environment, the blood must be thermally condi-tioned with the aid of special apparatus to preserve its viability. Research in this area is focused on the development of the requisite hardware which will produce minimal deleterious side effects.

10.   In clinical practice, heat transfer also plays a major role in the treatment of hyperthermia and hypothermia. Hyperthermia (higher than normal body temperature) is usually attendant to some other physiological disorder and is simply one of the ways in which the human body responds to the perturbation of its normal state. In such cases the prime concern from the clinical point of view is to treat first the cause of the hyperthermia itself. Similarly, in the case of hypothermia (lower than normal body temperature) the clinician must first treat the cause of the hypothermia and then the hypo-thermia itself. However, hypothermia is a little different from hyperthermia in that once the body temperature drops below approx-imately 85° F, the body loses the ability to regulate the body tempera-ture. Thus, the clinician must take steps to increase the body tem-perature above this limit as quickly as possible. Hypothermia also differs from hyperthermia in that it may be intentionally induced on either a local or general level in a given patient in order to take advantage of one of its side effects such as the reduction in metabo-lism or oxygen requirements. The major research effort in this area is devoted to developing techniques by which hyperthermia and hypo-thermia may be controlled either with the aid of extracorporeal pro-cedures or by producing the proper thermal environment for the organism.

11.   One final application of heat transfer in biomaterials that should
be mentioned here is waste heat rejection from artifical organs. If
artificial organs are to be implanted in humans for long periods of
time, they must be cyclic devices (in the thermodynamic sense) with
self-contained energy sources and energy sinks in order to satisfy
the second law of thermodynamics. If the human body itself is to be
used as the energy sink, careful attention must be given to the effects
of the rejected waste heat on the state of the organism. Current
research efforts in this area are devoted to the development of im-
plantable heat exchanger devices and to the investigation of the effects
of endogenous heat transfer on the physiological state of the organism.

THERMAL INTERACTIONS BETWEEN THE HUMAN BODY AND THE ENVIRONMENT

It is the purpose of the thermoregulatory system of the human body

to maintain a reasonably constant internal body temperature within

some fairly well defined limits. Although it is not the object of the

present work to describe the mechanisms by which the thermo-

regulatory system achieves this end, we do point out that in the main

these mechanisms involve changes in circulation, respiration, and

metabolism to counteract the effects of thermal interactions between

the human body and its environment. Clearly, the more severe the

interactions, the greater the human body in general and the thermo-

regulatory system in particular will be taxed. Thus in order to

minimize the demands placed upon the body, and hence enhance the

comfort of the individual, it is to our benefit to control these interactions as nearly as we can.

The nature of these interactions can be established by treating the human body as a control volume interacting with the environment across a control surface. The interactions are of two types, mechanical and thermal. While the mechanical interaction is certainly work transfer, the thermal interaction is not strictly heat transfer in the true thermodynamic sense. Nevertheless, in the steady state we can write (in the absence of any mass transfer across the control surface exclusive of respiration) [1:61]

$$M = H + W \qquad (1)$$

where M is the rate of energy production by metabolism, H is the rate of energy transfer by thermal interactions with the environment, and W is the rate at which the human being does work.

The rate at which the human being does work can be represented most conveniently by defining a mechanical efficiency, η, such that [1:61]

$$\eta = \frac{W}{M} \qquad (2)$$

and the value of η depends upon the nature of the physical activity in which the individual is engaged. Table 1 presents typical values of η for some common physical activities.

There are several mechanisms which contribute to the rate of

Table I.   Heat transfer data for common physical activities [1:61].[*]

Type of activity	Metabolic rate per unit body surface area M/A_{Du} kcal/m² hr	Estimated mechanical efficiency	Estimated relative velocity in "still" air m/s	Correction factors for effective radiation area f_{eff}
Seated, quiet	50	0	0	0.65
Seated, drafting	60	0	0-0.1	0.65
Seated, typing	70	0	0-0.1	0.65
Standing at attention	65	0	0	0.75
Standing, washing dishes	80	0-0.05	0-0.2	0.75
Shoemaker	100	0-0.10	0-0.2	0.65
Sweeping a bare floor (38 strokes/min)	100	0-0.05	0.2-0.5	0.75
Seated, heavy leg and arm movements (metal worker at a bench)	110	0-0.15	0.1-0.3	0.65
Walking about, moderate lifting or pushing (carpenter metal-worker, industrial painter)	140	0-0.10	0-0.9	0.75
Pick and shovel work, stone mason work	220	0-0.20	0-0.9	0.75
Walking on the level with the velocity (mph):				
2.0	100	0	0.9	0.75
2.5	120	0	1.1	0.75
3.0	130	0	1.3	0.75
3.5	160	0	1.6	0.75
4.0	190	0	1.8	0.75
5.0	290	0	2.2	0.75
Walking up a grade: Grade (%) / Velocity (mph)				
5 / 1	120	0.07	0.4	0.75
5 / 2	150	0.10	0.9	0.75
5 / 3	200	0.11	1.3	0.75
5 / 4	305	0.10	1.8	0.75
15 / 1	145	0.15	0.4	0.75
15 / 2	230	0.19	0.9	0.75
15 / 3	350	0.19	1.3	0.75
25 / 1	180	0.20	0.4	0.75
25 / 2	335	0.21	0.9	0.75

[*]Note: A_{Du} is the Dubois area, i.e., the body surface area of the nude body (meters²) [2:447] and f_{eff} is the ratio of the effective radiation area of the clothed body to the surface area of the clothed body.

energy transfer by thermal interactions with the environment.

Accordingly, we can write

$$H = D + S_W + R_W + R_d + Q \tag{3}$$

where D is the rate of energy transfer by diffusion of water vapor

through the skin, S_W is the rate of energy transfer by the secretion

of sweat, R_W is the rate of energy transfer associated with the

difference in water vapor content between the air expired and inspired

during respiration, R_d is the rate of energy transfer associated with

the difference in the enthalpy of the air expired and inspired during

respiration, and Q is the rate of energy transfer from the outer

surface of the clothing by radiation and convection.

By making certain assumptions regarding the physical processes

which contribute to equation (3), it is possible to relate each of the

terms appearing on the right-hand side to the physical parameters

that describe the state of the environment. For the details of this

procedure, the reader is referred to reference [1:61]. For the pur-

poses of the present work it is sufficient to simply list the relevant

expressions for each of the contributions to equation (3). For the

diffusion contribution

$$D = 0.35 \, A_{Du} \, (p_s - p_a) \quad \text{(kcal/hr)} \tag{4}$$

where p_a is the partial pressure (mm Hg) of water vapor in ambient air and p_s is the saturation pressure corresponding to the mean skin

temperature t_s ($^{\circ}$C) for an individual. From the thermodynamic pro-

perties of water in the temperature range $27^{\circ}C < t_s < 37^{\circ}C$, we can

write

$$p_s = 1.92\, t_s - 25.3 \qquad (mm\ Hg) \tag{5}$$

Because of variations in the physiology of individuals, the skin tem-

perature will vary from one individual to another even though they

are engaged in the same physical activity. Furthermore, skin tem-

perature can be related to that very subjective physical character-

istic, comfort. Although comfort is subjective and hence will vary

from one individual to another under a given set of physical condi-

tions, it is possible to average these interpretations over a large

population. The result of such averaging is a mean skin temperature,

$\overline{t}_s$, which will assure thermal comfort dependent the level of physical

activity. As reported in reference [1:61], the mean skin temperature

to be used in equation (5) is

$$\overline{t}_s = 35.7 - 0.032\, \frac{M}{A_{Du}}\, (1 - \eta) \qquad (^{\circ}C) \tag{6}$$

The sweat secretion term appearing in equation (3) is a function of

the level of physical activity as well as the physical parameters that

characterize the environment. Of course, even in a given situation,

there is still the question of thermal comfort which will vary from one individual to another. But again, as in the case of the mean skin temperature, we can average over a large population. The result of such averaging is a mean rate of energy transfer by sweating which will assure thermal comfort dependent upon the level of physical activity. As reported in reference [1:61], this mean rate of energy transfer by sweating is

$$\overline{S}_W = 0.42\, A_{Du} \left[\frac{M}{A_{Du}}\,(1-\eta)\ -50 \right] \qquad \text{(kcal/hr)} \qquad (7)$$

For the rate of energy transfer associated with the difference in water vapor content between the air expired and inspired during respiration,

$$R_W = 0.0023M\,(44 - p_a) \qquad \text{(kcal/hr)} \qquad (8)$$

where again the effects of thermal comfort have been included.

For the rate of energy transfer associated with the difference in the enthalpy of the air expired and inspired during respiration,

$$R_d = 0.0014\, M\,(34 - t_a) \qquad \text{(kcal/hr)} \qquad (9)$$

where t_a is the ambient air temperature ($^\circ$C) and the temperature of the expired air is assumed to be 34° C.

In calculating the rate of energy transfer from the outer surface of the clothing by radiation and convection, we must recognize that the energy being transferred by these two mechanisms first must have been conducted through the clothing. Using the familiar electrical

analogy for heat transfer by conduction [3, 98], we can write

$$Q = \frac{t_s - t_{cl}}{R} \tag{10}$$

where t_{cl} is the outer surface temperature of the clothing and R is

the resistance to heat transfer. For the sake of simplifying calcula-

tions Gagge, et al. [4:82] introduced a dimensionless quantity, I_{cl},

defined as the ratio of the thermal conduction resistance of a given

piece of clothing to the thermal conduction resistance of the average

business suit with underclothing. Thus in units of clothing

$$I_{cl} = \frac{R_{cl}}{R_{ave.\,bus.\,suit}} \quad (clo) \tag{11}$$

where

$$R_{ave.\,bus.\,suit} = 0.18\ m^2\ {}^{\circ}C\ hr/kcal \tag{12}$$

Then, combining equations (10), (11), and (12), we obtain

$$Q = A_{Du}\ \frac{t_s - t_{cl}}{0.18\ I_{cl}} \tag{13}$$

where the factors I_{cl} and f_{cl} (increase in body surface area due to

clothing) are tabulated in Table 2.

In the steady state, the energy transferred through the clothing by

conduction must ultimately be transferred to the environment by rad-

iation and convection.

Table 2.   Thermal Data for Various Clothing Ensembles

Clothing Ensemble	I_{cl} (clo)	f_{cl}
Nude	0	1.0
Light working ensemble: jockey shorts, wool socks, cotton work shirt and work trousers; shirt tail out, neck open	0.6	1.1
U.S. Army "fatigues", man's: lightweight underwear, cotton shirt and trousers, cushion sole socks, and combat boots	0.7	1.1
Typical American business suit	0.7 -1.0	1.1 -1.15
Light outdoor sportswear: cotton shirt, trousers, T-shirt, shorts, socks, shoes, and single ply poplin (cotton and dacron) jacket	0.9	1.15
Heavy traditional European business suit: light cotton underwear with long legs and sleeves, shirt, woolen socks, leather shoes, tweed suit with trousers and fully lined jacket and vest	1.2	1.15-1.2
U.S. Army standard cold-wet uniform: cotton-wool undershirt and drawers, wool and nylon flannel shirt, wind resistant, water repellent trousers and field coat, cloth mohair and wool coat liner, and wool socks	1.5 -2.0	1.3 -1.4
Heavy wool pile ensemble (polar weather suit)	3 - 4	1.3 -1.5

Thus

$$Q = Q_{rad.} + Q_{conv.} \tag{14}$$

where [1:61]

$$Q_{rad.} = 4.8 \times 10^{-8} A_{Du} f_{cl} f_{eff} \left[(T_d)^4 - (T_{rmt})^4 \right] \quad (kcal/hr) \tag{15}$$

where T_{cl} is the thermodynamic temperature of the outer surface of

the clothing (^{O}K) and T_{rmt} is the mean radiation temperature (^{O}K),

i.e., the uniform thermodynamic temperature of a black enclosure

which would give the same energy transfer by radiation from the

individual as the actual enclosure.

Also

$$Q_{conv.} = A_{Du} f_{cl} h_{conv.} (t_{cl} - t_a) \quad (kcal/hr) \tag{16}$$

where $h_{conv.}$ is the film coefficient for convective heat transfer and

is given by [1:61]

$$h_{conv.} = 2.05 (t_{cl} - t_a)^{1/4} \quad (kcal/m^2 \ hr \ ^{O}C) \tag{17}$$

if

$$2.05 (t_{cl} - t_a)^{1/4} > 10.4 \sqrt{v} \tag{18}$$

where v is the relative air velocity (m/sec) or

$$h_{conv.} = 10.4 \sqrt{v} \quad (kcal/m^2 hr \ ^{O}C) \tag{19}$$

if

$$2.05 \ (t_{cl} - t_a)^{1/4} < 10.4 \sqrt{v} \ . \tag{20}$$

The appropriate expressions for the various interactions between the human body and the environment can be evaluated and substituted into equation (1). The expression that results contains not only the physical characteristics of the environment but also the subjective judgement of thermal comfort (suitably averaged) of a given population (in the present case, college aged Americans) [1:61]. This expression, known as the comfort equation, must be satisfied as a necessary condition for thermal comfort. Because the solution of the comfort equation is quite laborious, it has been solved on a digital computer for a number of characteristic cases. For the numerical solutions and further details, the reader is referred to reference [1:61].

The foregoing discussion represents one way of characterizing the interactions between the human body and the environment. For further work in this area, the reader is referred to references [4:82] through [6].

THERMAL MODELING OF BIOMATERIALS AND THE MEASURE-
MENT OF THE THERMAL PROPERTIES OF BIOMATERIALS

The thermal modeling of biomaterials and the measurement of the thermal properties of biomaterials must be treated simultaneously since they are interdependent. While it is possible to discuss and analyze thermal models of biomaterials without making measurements of the various thermal properties, these analytical treatments are virtually useless unless numerical values can be obtained for the various solutions. Clearly such numerical results can be obtained only if the requisite thermal data are available. On the other hand, measurements of the thermal properties of biomaterials usually involve controlled thermal interactions between the biomaterial and the environment in a specific geometrical configuration. To obtain thermal property data from such measurements, the analytical expressions describing the thermal behavior of the biomaterial must be solved for the specified boundary conditions and geometry. These solutions are then combined with the experimental data to yield the thermal properties. Thus, the better the thermal model, the better the thermal properties; and the better the thermal properties, the better the models can be utilized.

In general, heat transfer processes in biomaterials are the result of complex interactions between tissue conduction, tissue perfusion (involving capillary blood flow as well as arterial and venous blood

flows), and tissue metabolism. To further complicate matters, bio-

materials are usually anisotropic and inhomogeneous. In view of

these complexities, two approaches have been employed to treat the

effects of conduction, perfusion, and metabolism while appropriate

zoning techniques have been used to treat the problems of anisotropy

and inhomogeneity. These two approaches are basically: (1) to

lump all non-conduction effects into the conductivity by suitably

defining effective thermal properties for the material or (2) to in-

clude separately the non-conduction effects in the appropriate govern-

ing equations for the model.

Obviously, the first approach is the simplest since the governing

equation for heat transfer becomes

$$\frac{\partial T}{\partial \tau} = \alpha_{eff} \nabla^2 T \tag{21}$$

where we have modeled the biomaterial as a homogeneous, isotropic

continuum with an "effective" thermal diffusivity, α_{eff}, ($\alpha_{eff} = k_{eff}/\rho c$

where k_{eff} is the "effective" thermal conductivity, ρ is the density of

the material, and c is its specific heat). T is the thermodynamic

temperature, and τ is the time. The advantage of this approach is

that equation (21) has been solved already for virtually every situation

imaginable [7]. However, the big disadvantage of the approach is that the resulting solution of equation (21) is applicable only to the specific

situation described by the initial conditions, boundary conditions,

geometry, perfusion rate, and metabolic heat generation rate. The

"effective" thermal diffusivity (or alternatively, the "effective"

thermal conductivity) <u>must</u> be evaluated for this specific situation

and cannot be applied to any other situation even though the initial

conditions, boundary conditions, and geometry may be identical.

That is, the resulting "effective" properties are functions of perfu-

sion rate (and possibly geometry and time) so that any variation in

tissue perfusion renders the result useless.

The second approach also treats the biomaterial as an isotropic,

homogeneous continuum permeated by a vascular network; however,

unlike the previous approach, the effects of tissue perfusion and

metabolism are accounted for separately in the governing equations.

In the most general case, we can distinguish between the arterial and

venous sides of the vascular network so that the equivalent form of

equation (21) for the biomaterial becomes

$$\rho c \frac{\partial T}{\partial \tau} = k \nabla^2 T + Q_{met} - c_b w_b (T - T_A) - U_A A_A (T - T_A) - U_V A_V (T - T_V) \quad (22)$$

where ρ is the density of the material, c is its specific heat, k is its

thermal conductivity, Q_{met} is the rate of metabolic heat generation

per unit volume, c_b is the specific heat of the blood, w_b is the mass rate of perfusion per unit volume of material, T_A is the thermodynamic temperature of the arterial blood, U_A is the overall heat transfer coefficient between the material and arteries, A_A is the surface area of the arteries per unit volume of material, U_V is the overall heat transfer coefficient between the material and the veins, and A_V is the surface area of the veins per unit volume of material.

Note that in equation (22) there are actually three unknowns: T, T_A, and T_V. Thus we require two additional equations in order to be able to solve equation (22). The necessary additional relations can be obtained by applying the first law of thermodynamics to the arterial and venous sides of the vascular network separately. The result for the arterial side is

$$\rho_b c_b \psi_A \frac{\partial T_A}{\partial \tau} + c_b \vec{G}_A \cdot \nabla T_A + U_A A_A (T_A - T) = 0 \tag{23}$$

where ρ_b is the density of the blood, ψ_A is the fraction of total material volume occupied by arterial blood, and $\vec{G}_A$ is the arterial blood mass flux vector per unit total area. For the venous side, we obtain

$$\rho_b c_b \psi_V \frac{\partial T_V}{\partial \tau} + c_b \vec{G}_V \cdot \nabla T_V + (c_b w_b + U_V A_V)(T_V - T) = 0 \tag{24}$$

where the individual venous terms in equation (24) are similar to corresponding arterial terms in equation (23). For discussion of the physical interpretation of the various terms in equations (22), (23), and (24) and their solutions, the reader is referred to reference [8]. The results of this approach are obviously more complete than the simplified approach; however, one must pay the price of increased mathematical complexity.

To illustrate the relative merits of the two approaches, let us consider a technique for measuring the thermal properties of bio-materials originally developed by Grayson [9:54] and subsequently improved upon by Chato [10:16, 11:395]. In essence, the method involves the implantation of an electrical resistance element in the biomaterial whose thermal properties are to be measured. The re-sistance element modeled as a sphere of radius R, has an electrical resistance which is a strong function of temperature so that the tem-perature of the element can be determined from measurements of its electrical resistance. Initially the element is in thermal equilibrium with the surrounding tissue at some temperature T_o. At some instant of time $\tau = 0$, the temperature of the element is increased to the uni-form value T_R by dissipating electrical energy in the element. The temperature T_R is maintained at this level for all subsequent times by means of a control system. If we assume that the blood enters the biomaterial at some arterial temperature, T_A, and is cooled or heat-ed in the capillaries to the local temperature, T, we can write the governing equation for the thermal behavior of the material in the form,

$$\rho c \frac{\partial T}{\partial \tau} = k \nabla^2 T - c_b w_b (T - T_A) + Q_{met} \tag{25}$$

For the initial and boundary conditions described above, equation (25) can be solved in spherical coordinates for the temperature distribution throughout the surrounding material. At the interface between the electrical resistance element and the surrounding material, the first law of thermodynamics requires that the heat transfer from the element to the surrounding material is related to the temperature gradient in the material evaluated at the interface. This heat transfer is of course due to the Joule heating in the electrical resistance element.

Let us consider two possibilities: (1) that the blood flow is occluded so that $w_b = 0$, and (2) that the blood flow is some fixed value $w_b \neq 0$. In the first case, the Joule heating can be shown to be [10:16]

$$q = [T_R - T_A] \{ 4\pi Rk + 4R^2 \sqrt{\pi k \rho c} \; \tau^{-1/2} \} \tag{26}$$

Thus a plot of the experimental values of q vs. $\tau^{-1/2}$ should be a straight line with the intercept at $\tau \to \infty$ proportional to k and the slope proportional to $\sqrt{k \rho c}$ (thermal inertia). From such a plot, the values of k and $\sqrt{k\rho c}$ can be determined. Alternatively, the second case the

Joule heating can be shown to be [10:16]

$$q = \left[T_R - T_A - \frac{Q_{met}}{w_b c_b} \right] \left\{ 4\pi Rk + \frac{4\sqrt{\pi}\, R^2 \sqrt{k\rho c}}{\sqrt{\tau}} \exp\left(-\frac{w_b c_b}{\rho c}\tau\right) \right.$$
$$\left. + 4\pi R^2 (k\, w_b c_b)^{1/2} \operatorname{erf} \left(\frac{w_b c_b \tau}{\rho c} \right)^{1/2} \right\} \tag{27}$$

Clearly in the second case the thermal properties cannot be evaluated

from the resulting solution quite so simply as in the first case. Note,

however, that in the steady state as $\tau \to \infty$ the heat transfer at the

interface becomes

$$q_{\tau \to \infty} = \left[T_R - T_A - \frac{Q_{met}}{w_b c_b} \right] \left\{ 4\pi Rk \left[1 + R(w_b c_b / k)^{1/2} \right] \right\} \tag{28}$$

If we compare the steady state form of equation (26) with equation

(28), we see that in the presence of blood flow we could define an

"effective" thermal conductivity, k_{eff}, such that

$$k_{eff}/k = 1 + R\,(w_b c_b / k)^{1/2}$$

Thus the "effective" thermal conductivity depends upon the geometry,

R, as well as the blood flow heat capacity rate, $w_b c_b$, and thermal

conductivity, k. Thus any attempt to use this "effective" thermal

conductivity for any situation except the one for which the measure-

ments were made is meaningless.

In the application of Chato's technique to the measurement of the thermal properties of tissue a number of difficulties were encountered. In particular it was found that the electrical resistance element had a finite thermal conductivity which resulted in thermal gradients within the element. Thus the temperature recorded for the element was some "averaged" temperature and not the uniform temperature as required by the model. Also the short time thermal response of the electrical resistance element suffered from thermal capacity effects so that the slope of the aforementioned plot was not constant. In addition, the effective size of the electrical resistance element was a function of the temperature difference between the element and the material far removed from the element. In summary, we find that the results for the thermal properties measured by this technique are somewhat less than satisfactory.

In recent years, several different techniques have been used to measure the thermal properties of biomaterials [12-15]. Most of these have been invasive in that they have required the implantation of either heat sinks or heat sources and thermocouples or other temperature sensors to measure the local temperature distributions. The most recent of these [14] has been considerably more successful than the others, primarily because of the simplicity of the experimental hardware. The technique employs a "thermocouple" of essentially infinite thermal conductivity with a heat capacity large enough so that the time history of the temperature of this thermocouple is easily recorded. When this thermocouple is imbedded in a biomaterial, it can be easily modeled as a lumped thermal mass in an infinite medium. The time behavior of the temperature predicted by this model is compared with the actual thermocouple response to determine the thermal properties of the surrounding medium. Although the numerical treatment of the experimental data is a bit laborious, the values obtained for the thermal properties appear quite good and have, in fact, shed some new light on the rate of blood flow in biomaterials. The major results of this effort are summarized in Table 3.

In spite of the success of the technique developed by Cooper and Trezek, it does suffer from the disadvantage that it, like most other methods, is invasive. That is, in order to utilize the technique it is necessary to insert a probe of some sort into a living biomaterial. As a result of this procedure, a track is formed which represents a disruption of the integrity of the biomaterial. More importantly, the area immediately adjacent to the probe experiences trauma which most likely alters the physiological state of the material. Cooper and

Table 3.   Thermal properties of <u>in-vitro</u> human organs[14].

Organ	Density (gm/cm^3)	Heat Capacity $(cal/gm^\circ C)$	% Water	Thermal Diffusivity (cm^2/sec)	Thermal Conductivity $(cal/cm\ sec^\circ C)$
Kidney	1.05	0.93	84	132×10^{-5}	130×10^{-5}
Heart	1.06	0.89	81	148×10^{-5}	140×10^{-5}
Spleen	1.05	0.89	80	138×10^{-5}	130×10^{-5}
Liver	1.05	0.86	77	150×10^{-5}	135×10^{-5}
Brain, White	1.04	0.86	71	134×10^{-5}	120×10^{-5}
Brain, Gray	1.05	0.88	83	143×10^{-5}	135×10^{-5}
Brain, Whole	1.05	0.88	78	138×10^{-5}	126×10^{-5}

Trezek [14] themselves report blood flow anomalies in the vicinity

of the probe which (in the opinion of this author) may affect the data

obtained in this manner. At the very least, the invasion of the bio-

material produces an artificial interface with its attendant contact

resistance which can have a marked effect on the experimental re-

sults. In fact, Chato [10:16] reports that contact resistances can be

so large in some cases that the experimental results are useless.

In order to overcome these deficiencies, it is necessary to develop

property measurement techniques which are non-invasive in nature.

One technique which holds promise in this regard is due to Hendler, et al. [15:177] . The method consists essentially of irradiating the surface of the biomaterial with infrared radiation and subsequently measuring the resulting surface temperature field as a function of time with the aid of an infrared detector. From this time-temperature histroy and a suitable model of the heat transfer processes in the material, the thermal inertia (kρ c) can be determined. In this manner, the integrity of the material is preserved, and the blood flow is virtually unaltered. The technique is enhanced by the fact that most biomaterials are nearly perfect absorbers of infrared radiation by virtue of their high water content. The only drawback of the technique is that it is sensitive to the heat transfer model employed in the calculation of the time-temperature history. Nevertheless, using this method, Hendler, et al. have determined the thermal inertia of human skin [15:177] with considerable precision. With a little ingenuity, it should be possible to extend this technique to the measurement of the thermal properties of other biomaterials.

Further progress in the measurement of the thermal properties of biomaterials must rely not only on the development of more sophisticated experimental techniques but also on the development of more precise models for the heat transfer processes in biomaterials. Recent advances in this area [16-19] have already shed new light on the heat transfer processes in biomaterials, but more work still remains to be done.

CRYOBIOLOGY AND CRYOSURGERY

It is beyond the scope of the present work to attempt to discuss all
the recent progress in these rapidly expanding fields of cryobiology
and cryosurgery, particularly in view of the many monographs recently published in this area [20-24]. However, in view of the role
heat transfer plays in the freezing processes in these applications,
it is worthwhile to mention here some of the more significant aspects
of cryobiology and cryosurgery.

Generally speaking, the focus of cryobiology is the successful preservation and storage of biomaterials (organs and tissues) by freezing. Although considerable effort has been expended on this problem,
success (as measured in terms of the clinical application of the technique) has been limited at present to relatively few biomaterials,
most notably blood, sperm, skin, cornea, and certain tissue cultures.
Primarily as a matter of convenience, cryobiological research has
concentrated on studies of the effects of temperature on the _in vitro_
behavior of cell suspensions. Briefly, the major findings of these
studies are:

1. The biochemical and physical processes by which biomaterials
sustain life are affected to varying degrees by temperature, but in
general, if low temperatures are successful in arresting these processes, then the lower the temperature for storage the better. That
is, for purposes of preserving biomaterials $-196^{\circ}C$ (the boiling
point of nitrogen at one atmosphere pressure) is preferable to $-79^{\circ}C$
(the saturation temperature of carbon dioxide at one atmosphere
pressure).

2. In terms of post-thaw survival after the freeze-thaw cycle, the
time rate of change of temperature is probably the single most important physical parameter. In particular, the time rate of change of
temperature prior to the appearance of the solid phase during the
cooling process is probably more important than the time rate of
change of temperature during the thawing process.

On the basis of this experience, data for the survival of cells subjected to freezing and thawing process is usually presented in terms
of the fraction of cells surviving as a function of the cooling velocity,
i. e., the time rate of change of temperature prior to the appearance
of the solid phase. Implicit in this approach is the tacit assumption
that the specimen was thawed at a rate sufficient to minimize damage
during this part of the freeze-thaw cycle. (Typically, biomaterials
are less sensitive to warming velocities than to cooling velocities.)
Fig. 1 illustrates the sort of "survival signatures" that result from
such studies. It is worth noting in Fig. 1 that the individual survival
signatures vary with respect to "skewness", cooling velocity for
maximum survival, maximum fraction of cells surviving, and "bandwidth" depending upon the cell type. There is no unique cooling velocity that will guarantee survival independent of cell type.

Although there are many physio-chemical changes that occur in a
cell suspension during freezing, it is generally agreed (at least for
the purposes of the present work) that the damage experienced by the
cell can be traced to two mechanisms:

1. The concentration of solutes in the extracellular fluid increases
as water is frozen out of solution. In an attempt to maintain osmotic
equilibrium across the cell membrane, water leaves the cell resulting in a decrease in cell volume as well as an increase in electrolyte

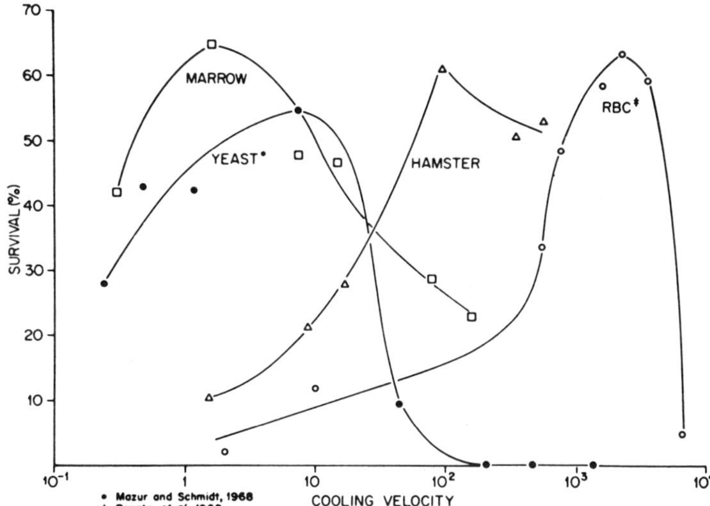

Figure 1. Survival signatures of selected biomaterials [25:69].

concentration. The high concentration of solutes is believed to
damage the cell membrane through the action of electrolytes on the
lipoproteins that make up the cell membrane. If we assume that
thermodynamic equilibrium prevails in the solution, the concentration
of solutes is a function of temperature but not the time rate of change
of temperature. However, the interaction between the electrolytes
and the lipoproteins is a rate process which apparently can be
"frozen" if the temperature is low enough. Thus there is a competition between the rate processes of membrane damage and heat transfer such that some critical cooling velocity exists beyond which no
cell damage occurs by this mechanism.

2. As described above, during freezing the cell attempts to maintain
osmotic equilibrium between the intracellular contents and the extracellular fluid by the transfer of water across the cell membrane.
This equilibration process is a rate process which depends in part
upon the permeability of the cell membrane to water. Furthermore,
there is a competition between this rate process and the rate process
of heat transfer such that there exists a critical cooling velocity
above which the rate of heat transfer exceeds the rate of water transfer. Thus water is trapped inside the cell in the form of ice. By
some mechanism, which at the present time is unclear, this intracellular ice damages the cell. There is some experimental evidence
to indicate that the cell membrane is somehow involved in the damage
process; however, the precise nature of the involvement is unknown

at the present time. For a more thorough discussion of the significance of intracellular ice on the survival of frozen biomaterials, the
reader should consult reference [26].

The hypothesis of cellular damage via these two mechanisms is
supported by the experimental evidence that has been accumulated to
date. It is an experimentally established fact that low cooling velocities favor the formation of extracellular ice while high cooling velocities favor the formation of intracellular ice. Thus as shown in Fig. 2,
we can construct a hypothetical, composite survival signature from
the individual survival signatures associated with each of these injury
mechanisms. Note that the general behavior of the composite survival signature is consistent with the experimentally determined survival
signatures of Fig. 1.

The nature of survival signatures can be best summarized by the
generalized survival signature of Fig. 3. Note that the values shown
for the cooling velocities are not to be interpreted in an absolute
sense but should be interpreted as representative of the relationship
between cooling velocities and cellular damage.

In an effort to quantify the terms "high cooling velocity" and "low
cooling velocity", Mazur [29:347] has proposed an analytical model to
describe the aforementioned physio-chemical processes that occur
during freezing. While it is beyond the scope of the present work to
describe the details of the analysis, we do cite the final result which
is a differential equation describing the volume of water in the cell as
a function of the thermodynamic temperature.

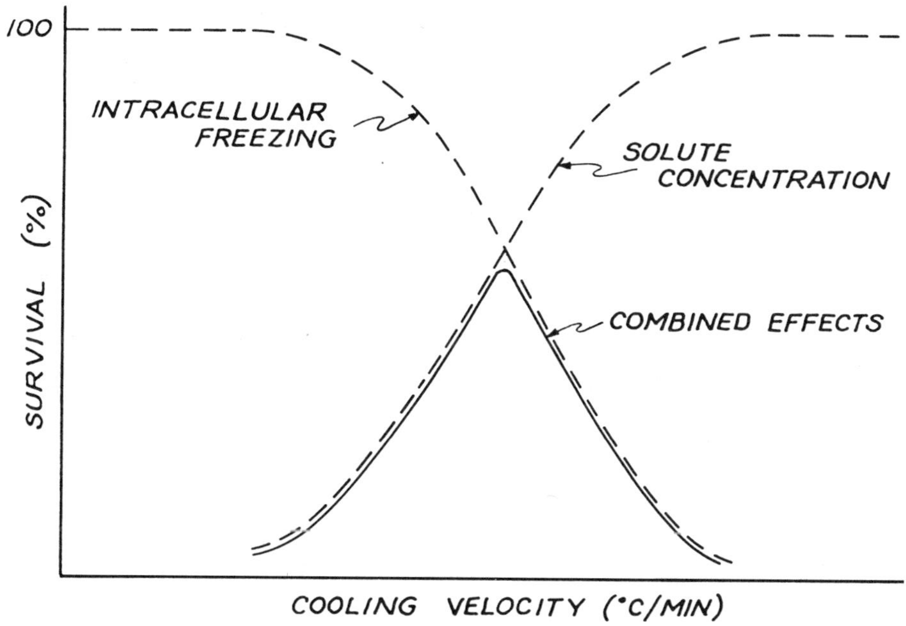

Figure 2. Hypothetical, composite survival signature based on two injury mechanism model [27:213].

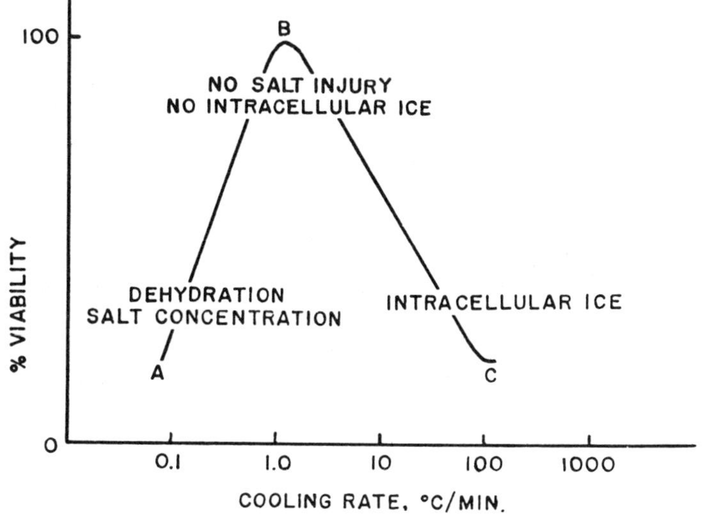

Figure 3. Generalized survival signature showing domains of freezing injury [28].

$$\frac{L_f A k_g}{B v_1^o} = T e^{b(T_g - T)} \frac{d^2 V}{dT^2} - \left[(bT + 1) e^{b(T_g - T)} - \frac{A R k_g n_2}{B(V + n_2 v_1^o)} \frac{T^2}{V} \right] \frac{dV}{dT}$$

(30)

where k_g is the permeability of the membrane to water at temperature T_g, b is the temperature coefficient of membrane permeability, n_2 is the number of osmoles of solute in the cell, v_1^o is the molar volume of pure water, A is the surface area of the cell membrane, B is the cooling velocity, L_f is the molar heat of fusion of ice, R is the gas constant, T is the thermodynamic temperature, T_g is the thermodynamic temperature at which the permeability coefficient is defined, and V is the volume of water in the cell. This equation cannot be solved analytically in this form, but Mazur has obtained numerical solutions on a digital computer for several typical cell types. More recently, Ling and Tien [30] modified the equation so that it can be expressed in non-dimensional terms. They have further defined certain limiting conditions, corresponding to particular physical circumstances, for which analytical solutions can be obtained. While the mathematics of the model has been worked out in detail, more experimental work remains to be done to test the underlying physical assumptions of the model.

In an effort to minimize the injury to cells during freezing, considerable research has been devoted to the study of the protection afforded cells by the addition of certain compounds known as cryoprotectants or cryophylactics. These compounds, which cover an

incredible spectrum of chemical properties range, from glycerol and
DMSO, to low molecular weight sugars such as sucrose and dextrose
to high molecular weight polymers such as polyvinylpyrrolidone (PVP)
and polyethylene glycol. These compounds fall into two broad categories: substances which penetrate the cell membrane and substances
which do not penetrate the cell membrane. It was originally felt that
the compound must penetrate the cell membrane to be effective, but
significant improvements in cell survival have been obtained with
additives such as the sugars and the polymers which do not penetrate
the membrane. As shown in Fig. 4, these compounds modify the unprotected survival signature very markedly, but the primary advantages of these compounds are that (1) they tend to "smear out" the
survival signature and render the cell less sensitive to cooling
velocity, i.e. the "bandwidth" of cooling velocities that favor survival
is broader than in the unprotected case, and (2) they tend to shift the
cooling velocity for maximum survival to lower cooling velocities
which can be more readily produced in clinical practice.

In spite of the considerable data available on these compounds, the
precise mechanisms by which they protect biomaterials during freezing are not well understood. It is generally believed that compounds
such as glycerol and DMSO which penetrate the cells and afford their
greatest protection at low cooling velocities, reduce the hypertonicity
of the extracellular fluid during freezing and therefore minimize cell
dehydration. On the other hand, compounds such as sucrose, dex-

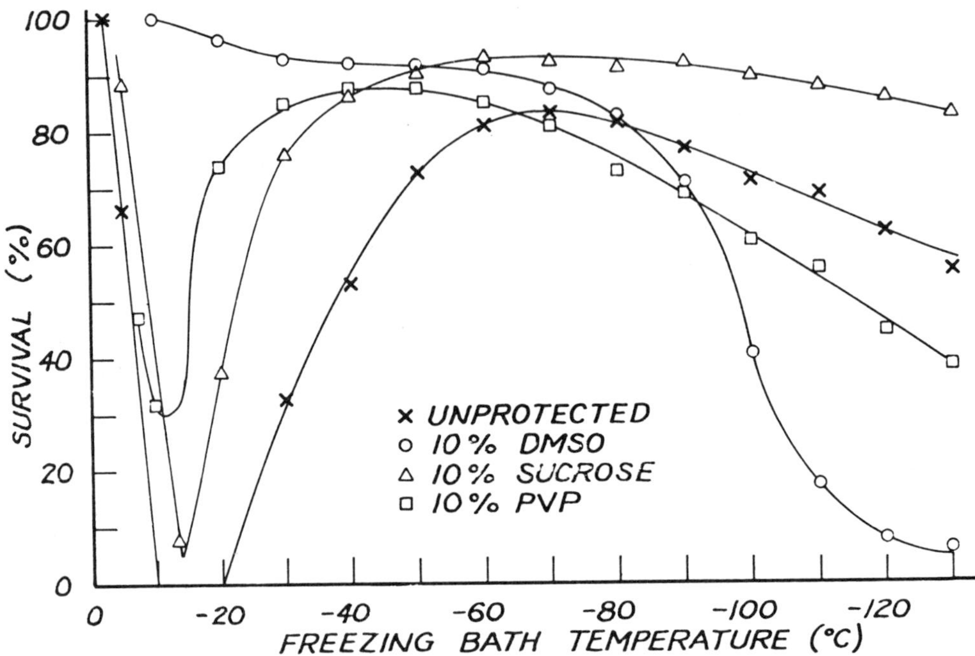

Figure 4.   Survival signatures for human erythrocytes frozen with different cryophylactic compounds [21].

trose, PVP, and polyethylene glycol which do not penetrate the cells and afford their greatest protection at high cooling velocities are generally believed to influence ice crystal growth or protect the membrane in some way. Although these compounds do protect bio-materials from freezing injury, they are not the panaceas that they might appear to be. They introduce another whole class of new prob-lems which may sometimes be more complex than the freezing injury problem itself. These compounds usually alter the tonicity of the cell suspension medium and in some cases may produce undesirable osmotic effects. Therefore they must be added to the cell medium under conditions which will minimize these effects. Furthermore, these compounds often must be used in concentrations that the human body cannot tolerate. Thus, special techniques must be developed for the removal of the protective compound before the cells can be used in the body. These removal or washing processes must also minimize attendant, undesirable osmotic effects.

Up to this point, we have discussed only the problems associated with the cryopreservation of cell suspensions without any consideration for the problems associated with the cryopreservation of a large volume of a biomaterial, such as a human organ. The omission has been intentional since these biomaterials must cope not only with precisely the same problems of the cell suspension, but also a special set of problems all their own. These problems arise primarily from the geometry and physical structure of the organ itself. In a system of any size, temperature gradients are a natural consequence of any heat transfer process. It follows, then, that any attempt at the cryopreservation of a biomaterial will result in the biomaterial being subjected to a spectrum of cooling velocities, the breadth of which will depend upon the thermal properties of the biomaterial. Furthermore, the biomaterial itself will most likely be made up of a number of different cell types, each with its own special survival signature. The net result of this whole situation is that some components of the biomaterial will be destroyed by the cryopreservation process while others will survive the process. We can best illustrate this point by the discussion of an example in the context of cryosurgery, but before we present the example, we wish to discuss first some general aspects of cryosurgery.

The objective in cryosurgery is to produce localized necrosis in biomaterials. The success of these procedures clearly depends upon the heat transfer characteristics of the material. Since the biomaterial is made up of cells, the previous discussion of survival signatures is just as relevant here as it was in the case of cryopreservation, except in the present context the emphasis is on cell death, not survival. To achieve the desired result, we must cool the biomaterial locally at a rate which guarantees cell death. The particular local cooling velocity to be used will depend upon the cell type. Thus, as in the case of the cryopreservation of bulk biomaterials, cryosurgery is hampered by the fact that the material is composed of several cell types, each with its own survival signature which must be respected. Here again, the problem is complicated further by the geometry of the material [31:316, 32:183].

For the purposes of illustration, consider the example of a typical cryosurgical probe discussed by Cooper and Trezek [32:183]. As shown in Fig. 5, the probe diameter is 3.5mm and its surface temperature is maintained at -125°C by a suitable control system. The cryoprobe is inserted into the tissue and the freezing procedure is commenced. If we assume that we can represent the surrounding tissue as a continuous, homogeneous medium, its response to the

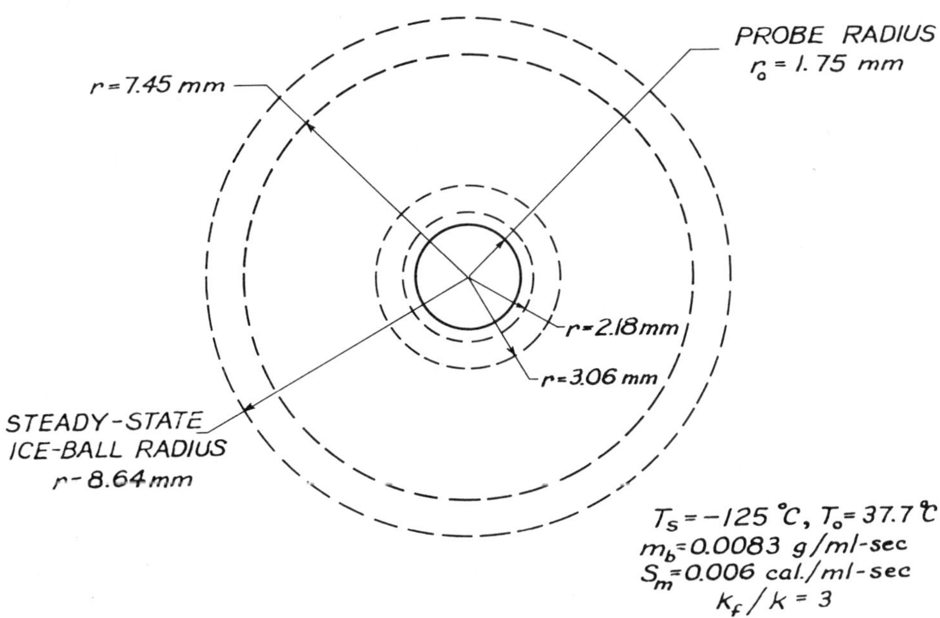

Figure 5.   Spherical cryoprobe and ice ball.

cryoprobe can be described in terms of a model and associated part-

ial differential equations similar to equation (25). After some period

of time (which is in fact infinitely long), the freezing process achieves

a steady-state condition and the mass of tissue in the frozen phase is

all contained within a sphere of known radius. The size of this

"steady-state ice ball" as calculated from the model is governed by

several parameters, particularly the thermal conductivity, blood

mass-flow rate, and metabolic heat generation rate of the surround-

ing tissue. For our purposes we have chosen values for these param-

eters typical of brain tissue.

Let us consider three particular locations within the steady-state ice ball: (1) Very close to the probe surface, (2) some distance away from the probe surface, and (3) near the steady-state ice ball radius. From the solutions of the temperature field, we can calculate the cooling velocities obtained at these locations throughout the period of time required to attain steady-state. These results are shown in Fig. 6 where we have also illustrated the significance of the survival signature. If, for example, the survival signature of the tissue in question is as shown in the uppermost signature, necrosis is a certainty for all the tissue contained within the steady-state ice ball. However, if the pertinent survival signature is as depected in the lower signature, a substantial fraction of the cells within the ice ball will survive and the cryosurgical procedure is ineffective. Thus with the aid of the technique described, the probable success of the cryosurgical procedure (or alternatively, reversible freezing process) can be predicted a priori provided the thermal properties, namely thermal conductivity, tissue perfusion, and tissue metabolism, are known.

The preceding illustration also serves to point out another fact which is particularly significant in the case of cryopreservation of biomaterials. If the survival signature is such that the model predicts that the freezing process virtually guarantees necrosis, how can we modify it with the aid of cryophylactic agents to produce more favorable results? The answer to this question can be found only

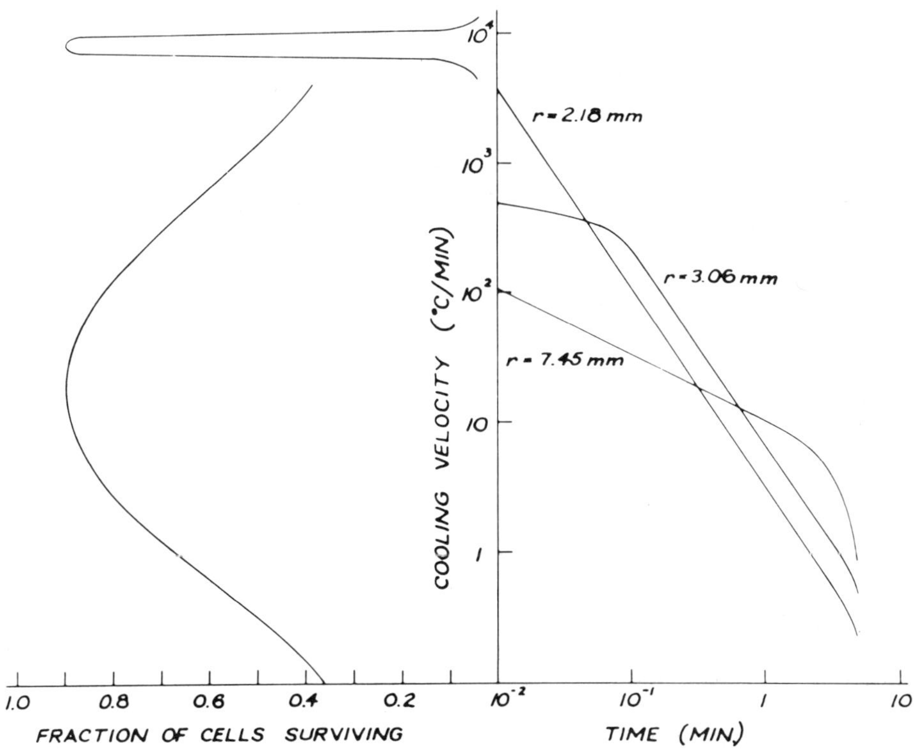

Figure 6.   Cooling velocities and survival signatures for spherical cryoprobe.

when the mechanisms of freezing injury are fully understood.

THERMOGRAPHY

As a final example of the role heat transfer plays in biomaterials,

we describe briefly a diagnostic procedure, known as thermography,

which is possible solely because of the heat transfer processes that

occur in biomaterials.   Basically, the procedure consists of making

a photographic analog of the thermal state of the biomaterial.   Biomaterials, like all physical objects that have temperature, emit electromagnetic radiation over a spectrum of wavelengths. In principle, the spectrum is infinite in extent, but in actual practice, only the radiation of a narrow portion of the spectrum is of interest for diagnostic purposes. The intensity of the emitted radiation depends primarily upon the temperature of the material, its radiation characteristics, and the direction in which it is emitted.

The essential elements of a device which samples this radiation for diagnostic purposes are: (1) a lens system for focussing the incoming radiation, (2) a detector which converts the focussed radiation into an electrical signal, (3) an electronics package which processes the electrical signal for imaging or display purposes, and (4) an imaging or display unit which displays as the processed data as the output of the device. As an example of such a device, consider the Barnes thermograph shown in Fig. 7 taken from reference [33:94]. Typically, the detector converts this incoming radiation to an electrical signal by some physical process such as a thermoelectric effect or a photoconductive effect depending upon the particular type of detector. This electrical signal is then suitably amplified to modulate the intensity of a light source in direct proportion to the intensity of the radiation incident on the detector. The modulated light is then directed to a film plate where it produces an image. By means of a scanning technique, the image recorded on the film will be essentially a photograph of the subject except that the relative density of

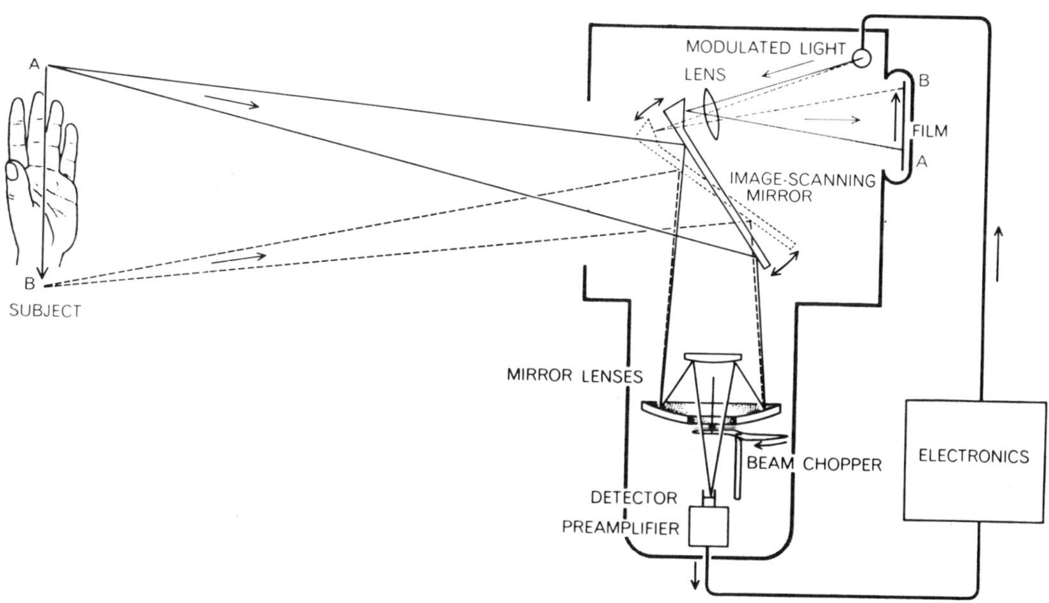

Figure 7. Schematic diagram of Barnes thermograph [33:94].

the image on the negative is proportional to the intensity of radiation
to which the detector is sensitive, rather than to the intensity of visible
light reflected from the subject.

Because of the magnitudes of their thermodynamic temperatures,
biomaterials tend to emit more intense radiation in the wavelength
region 4-20 μ (infrared) than in any other region. The human body
in particular is a very efficient emitter of radiation at these wavelengths. For this reason, devices of this type are designed to be
sensitive to radiation at these wavelengths; thus the density of the

image recorded on the film will be proportional to the surface temperature of the subject, and the resulting image is known as a thermogram. The local densities of the images recorded on the film plate are such that regions of high temperature appear whiter than regions of low temperature on the black and white positives produced from these negatives. Thus the positive record is a thermal map of the subject with local temperatures depicted on a gray scale. Temperature anomalies arising from a variety of sources such as metabolic and circulatory abnormalities would then appear as anomalies in the relative grayness of the image. For example, Fig. 8 is a thermogram of a patient with a thyroid tumor. Because of the high metabolic heat generation rate of the tumor, the region of tissue surrounding the tumor will be at a higher temperature than normal. In this case, the perturbation of the temperature field extends to the surface so that the infrared radiation emitted by the surface is higher than normal. Thus a white area appears over the location of the thyroid.

Fig. 8 also illustrates one of the deficiencies of thermography. As in the case of interpreting X-ray photographs, some experience is required both to detect the temperature anomalies and to diagnose the anomaly once it has been detected. For thermographs currently in use, temperature resolution typically is on the order of 1°C. To help sharpen the resolution, objects of known emissivity and temperature can be included with the subject in the field of view, but the operator still must rely on his experience to know what temperature

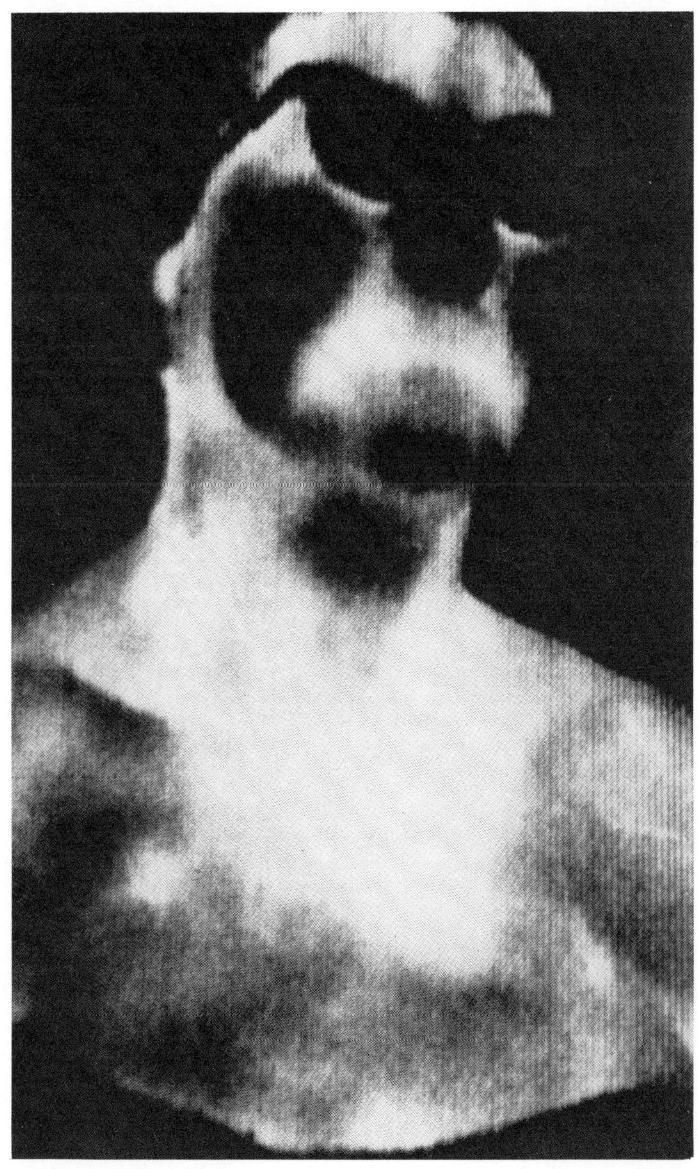

Figure 8. Thermogram of a patient with a thyroid tumor [33:94].

distributions are normal for a given region of the body. As improvements in data processing equipment are developed, the burden on the operator should be reduced considerably. For example, image enhancement[1] techniques are developed for photographic missions in the space program should prove invaluable to the operator. With the proper data processing equipment, the operator should be able to modify the degree of image enhancement on-line and thereby speed up the detection phase of operation. Detection can also take advantage of the symmetry of the human body by making contralateral comparisons as shown in Fig. 9.

The uppermost illustration in Fig. 9 is the thermogram of the chest of the patient. The tumor is located at X on the left-hand side as indicated by the letter L. A densitometer[2], or some other suitable device, can be used to scan the image. Four such scans were used in this example. Scan A was a scan of elements at known temperatures used to calibrate the device. Scans B, C, and D were scans at various locations on the patient. Note the symmetry of scans B and

---

[1]Image enhancement is a data processing procedure in which certain elements of a record are adjusted according to some preselected formula. In photography, the film record is divided up into decrete elements, or bits, and the average density of the image is intensified or de-intensified to improve the relative contrast between elements.

[2]A densitometer is a photoelectric device used to measure the density of the image on a photographic negative. Visible light is transmitted through a small region on the negative to a photoelectric element. The more light that passes through the negative, the greater will be the magnitude of the output signal of the device.

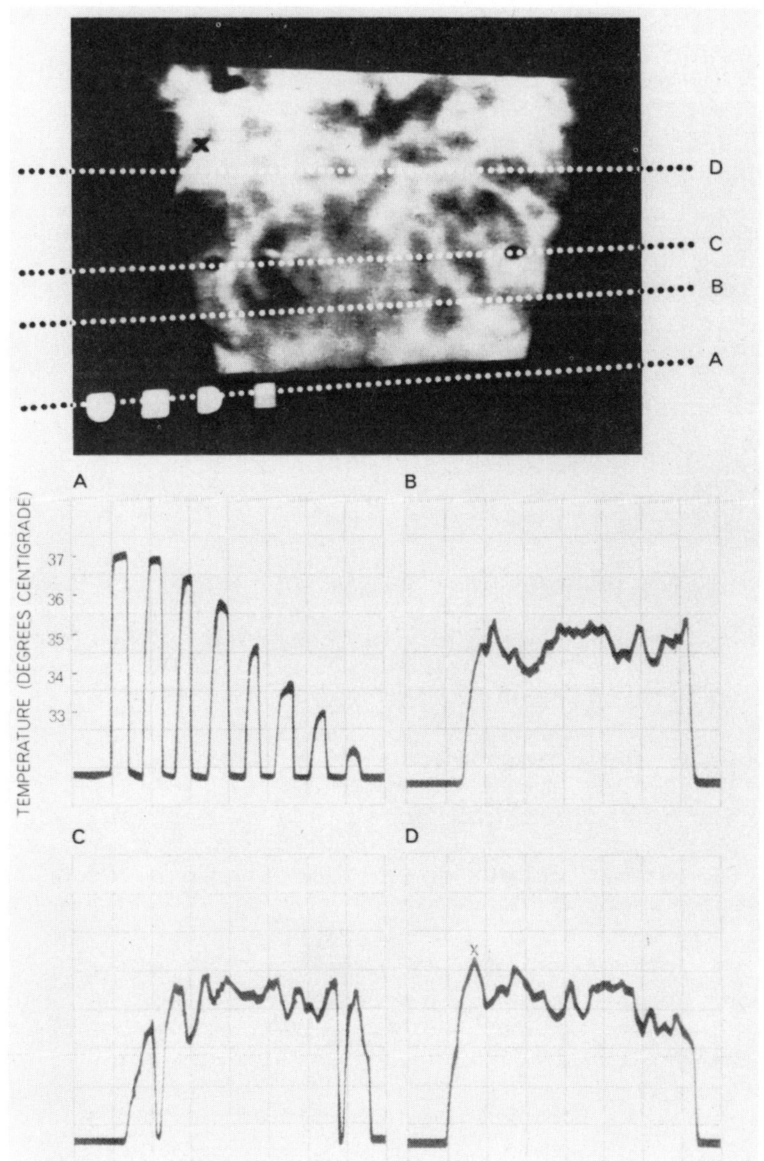

Figure 9.   Detection of breast tumor by contralaterial comparisons
in thermogram [33:94].

C and the lack of symmetry in scan D. The temperature spike at location X is typical of breast tumors but can be diagnosed as such only if the operator has the experience required to make such a diagnosis.

The potential of thermography as a non-destructive diagnostic device is only just beginning to be realized by those who are in a position to utilize it. As technology expands in this area, more rapid and more sensitive instruments with higher resolution are bound to develop; however, one aspect of this technique that is in desperate need of assistance from the physical scientist, is in the development of diagnostic aids. More specifically, the thermal modeling techniques described earlier in the present work can be utilized to establish the relationship between temperature anomalies recorded on the thermogram and the magnitude and location of the physiological defect. For example, what must be the extent of a tumor or other local metabolic disorder under given curcumstances of tissue perfusion and thermal properties before it will appear on the thermogram? What influence does the depth of the metabolic disorder play in its detection? Similar questions remain to be answered for circulatory defects as well. Until these questions and others like them are answered by the physical scientist, the potential of thermography can never be fully realized. Hopefully, thermal modeling techniques, together with high speed data processing devices, can provide the necessary answers.

For a more thorough and complete discussion of the many aspects of thermography, the reader is referred to reference [34:1].

CONCLUSION

Heat transfer in biomaterials can play a major role in the diagnosis and therapy of a broad spectrum of clinical disorders, as well as increasing our basic understanding of human physiology. The constraints of time and space have prevented us from discussing all of these applications or even from discussing a few of them in the depth that is warranted. Rather, the foregoing discussion represents only a brief introduction into an area of physical science that is yet to receive the full measure of attention that it deserves.

ACKNOWLEDGMENTS

The author would like to acknowledge the work of Professor B. B. Mikic, Department of Mechanical Engineering, M. I. T. in the field of heat transfer in biomaterials. His thoughtful and illuminating discussions and lectures on this subject, particularly in the area of thermal modeling, have contributed in no small way to the present work.

REFERENCES

[1]   Fanger, P.O. , R. G. Nevins, and P.E. McNall. 1968. Predicted and Measured Heat Losses and Thermal Comfort Conditions
for Human Beings. In Thermal Problems in Biotechnology, pp. 61-82.
New York: American Soceity of Mechanical Engineers.

[2]   Hardy, J. D. and E. F. Dubois. 1938. Basal Metabolism, Radiation, Convection, and Vaporization at Temperatures of 22 to 35°C.
J. Nutrion. 15:477.

[3]   Rohsenow, W. M. and H. Y. Choi. 1961. Heat, Mass, and
Momentum Transfer, pp. 98-99. New York: Prentice-Hall, Inc.

[4]   Buchberg, H. and C. B. Harrah. 1968. Conduction Cooling of the
Human Body - A Biothermal Analysis. In Thermal Problems in Bio-
technology, pp. 82-96. New York: American Soceity of Mechanical
Engineers.

[5]   Drake, R. M. , J. E. Funk, and J. B. Moegling. 1968. Sensible
Heat Transfer in the Gemini and Apollo Space Suits. In Thermal
Problems in Biotechnology, pp. 96-112. New York: American Society
of Mechanical Engineers.

[6]   Stoll, A. M. 1967. Heat Transfer in Biotechnology. In Advances
in Heat Transfer, Vol. IV, J. P. Hartnett and T. F. Irvine, Jr. , eds.
New York: Academic Press.

[7]   Carslaw, H. S. and J. C. Jaeger. 1959. Conduction of Heat in
Solids. 2nd ed. Oxford University Press.

[8]   Keller, H. K. and L. Seiler, Jr. 1969. An Analysis of Peripheral Heat Transfer in Humans. Memorandum Department of Chemical
Engineering, University of Minnesota.

[9]   Grayson, J. 1952. Internal Calorimetry in the Determination of
Thermal Conductivity and Blood Flow. J. of Physiology. 118:54.

[10]   Chato, J. C. 1968. A Method for the Measurement of the Thermal Properties of Biological Materials. In Thermal Problems in
Biotechnology, pp. 16-25. New York: American Society of Mechanical Engineers.

[11]   ___, 1969. Heat Transfer in Bioengineering. In Lectures on
Advanced Heat Transfer, ed. B. T. Chao, pp. 395-408. Urbana,
Illinois: University of Illinois Press.

[12]   Trezek, G.J., D.J. Jewett, and T.E. Cooper. 1968. Measurements of In-Vivo Thermal Diffusivity of Cat Brain. In Proceedings of the Seventh Conference on Thermal Conductivity, D.R. Flynn and B.A. Peavy, Jr., eds., pp. 749-754. Washington, D.C.: National Bureau of Standards.

[13]   Trezek, G.J. and T.E. Cooper. 1968. Analytical Determination of Cylindrical Source Temperature Fields and their Relation to Thermal Diffusivity of Brain Tissue. In Thermal Problems in Biotechnology, pp. 1-15. New York: American Society of Mechanical Engineers.

[14]   Cooper, T.E. and G.J. Trezek. 1970. A Probe Technique for Determining the Thermal Conductivity of Tissue. ASME Paper No. 70-WA/HT-18.

[15]   Hendler, E., R. Crosbie, and J.D. Hardy. 1958. Measurement of Heating of the Skin during Exposure to Infrared Radiation. J. of Applied Physiology. 12:177.

[16]   Shitzer, A. and J.C. Chato. 1970. Analytical Modeling of the Thermal Beahvior of Living Human Tissue. Paper Cu 3.9 presented at the Fourth International Heat Transfer Conference, Paris-Versailles.

[17]   ___, 1971. Steady-State Temperature Distribution in Living Tissue Modeled as Cylindrical Shells. ASME Paper No. 71-WA/HT-34.

[18]   ___, 1971. Analytical Solutions to the Problem of Transient Heat Transfer in Living Tissue. ASME Paper No. 71-WA/HT-36.

[19]   Chato, J.C. and A. Shitzer. 1971. Thermal Modeling of the Human Body - Further Solutions of the Steady State Heat Equation. AIAA Journal. 9:856.

[20]   Meryman, H.T., ed. 1966. Cryobiology. New York: Academic Press.

[21]   Smith, A.U., ed. 1970. Current Trends in Cryobiology. Plenum Press.

[22]   Wolstenholme, G.E.W. and O'Connor, M. eds. 1970. The Frozen Cell. London: J.A. Churchill.

[23]   Rand, R.W., Rinfret, A.P., and von Leden, H. eds. 1968. Cryosurgery. Springfield, Illinois: C.C. Thomas Publishers.

[24]   Zacarian, S. A. 1968. Cryosurgery of Skin Cancer. Springfield, Illinois: C. C. Thomas Publishers.

[25]   Mazur, P. et al. 1970. Interactions of Cooling Rate, Warming Rate and Protective Additive on the Survival of Frozen Mammalian Cells. In The Frozen Cell, Wolstenholm, G. E. W. and O'Connor, M. eds., pp. 69-89. London: J. and A. Churchill.

[26]   Diller, K., E. G. Cravalho, and C. E. Huggins. 1972. The Significance of Intracellular Ice on the Survival of Frozen Biomaterials. To be presented at the AIAA Seventh Thermophysics Conference in San Antonio, Texas.

[27]   Mazur, P. 1966. Physical and Chemical Basis of Injury in Single-celled Micro-organisms Subjected to Freezing and Thawing. In Cryobiology, ed. H. T. Meryman, pp. 213-315. New York: Academic Press.

[28]   Rowe, A. 1966. The Significance of the Aqueous-Ice Phase Transformation during Controlled Rate Cooling of Biological Specimens. In Cellular Injury and Resistance in Freezing Organisms, ed. E. Asahina. Proceedings of the International Conference on Low Temperature Science, Sapporo, Japan.

[29]   Mazur, P. 1963. Kinetics of Water Loss from Cells at Subzero Temperatures and the Likelihood of Intracellular Freezing. J. Gen. Physiology. 47:347.

[30]   Ling, G. R. and C. L. Tien. 1969. Analysis of Cell Freezing and Dehydration. ASME Paper No. 69-WA/HT-31.

[31]   Barron, R. F. 1968. Heat Transfer Problems in Cryosurgery. J. Cryosurgery. 1:316.

[32]   Cooper, T. E. and G. J. Trezek. 1970. Rate of Lesion Growth around Spherical and Cylindrical Cryoprobes. J. Cryobiology. 7:183.

[33]   Gershon-Cohen, J. 1967. Medical Thermography. Scientific American. 216:94.

[34]   Wipple, H. E. ed. 1964. Thermography and Its Clinical Applications. Annals of the New York Academy of Sciences. Vol. 121, art. 1, pp. 1-306.

CHAPTER 12

ULTRASOUND IN BIOLOGY AND MEDICINE

Padmakar P. Lele

Since its introduction into medicine some 25 years ago ultrasonics

has made great strides in both the diagnostic and therapeutic fields

and is being used to an ever increasing extent in clinical medicine.

The reasons for this are two-fold; first, it often provides informa-

tion regarding tissue structure that is not obtainable during the life

of the patient by any other diagnostic technique, and second, conven-

tional radiological techniques have proven to be more hazardous than

was believed previously. Thus, for instance, using rather simple

pulse-echo techniques, in ophthalmology it is possible to measure

the dimensions of the various components of the eye more accurately

than is possible using light and it is possible to detect conditions such

as detachment of the retina even when obscured by a hemorrhage in

structures lying in front of it; in cardiology it is possible to diagnose

accumulation of fluid in the pericardial sac or the structural and

functional state of several valves of the heart; and in obstetrics it is

possible to diagnose pregnancy in the presence of fibroid tumors of

the uterus. Thus, in fact, virtually no organ is inaccessible to ultrasonic examination. In addition to the pulse-echo techniques, ultrasonic Doppler methods are also used extensively for detection of movement such as pulsations in blood vessels. This enables detection of pregnancy as soon as the fetal heart is accessible to ultrasonic examination; or diagnosis of thrombosis (blockage) of veins deep in the leg.

In all these applications, as far as is known, no irreversible or permanent alterations are produced in the tissues as a result of their irradiation with ultrasound. On the contrary, it is the beam of ultrasound that is modified by its passage through the tissues and the type and extent of this alteration yields information on organ struc-ture. At higher ultrasonic intensities (and dosages) however, tissues cannot completely recover from the effects of ultrasonic irradiation and permanent structural alterations result. Ultrasound at such intensity levels is therefore used for therapeutic purposes such as in the treatment of Ménière's Disease where ultrasonic surgery is the treatment of choice and can alleviate the illness without producing deafness. Another such application is the surgery of the cataract of the eye. For these diagnostic and therapeutic applications ultrasound is used at frequencies of about 1 to 10MHz. At lower frequencies it is used extensively for disruption of cell walls for extraction of thermolabile intracellular ingredients. A discussion of all the current

applications, their underlying physical principles, and instrumenta-

tion can be found in some of the general texts [1-4] and specialized

publications [5-8]. A brief outline is presented here of some of the

work currently being conducted in the author's laboratory. We deal

first with the development of focused ultrasound as a surgical tool

for clinical and research use.

An ideal surgical tool should have the following characteristics:

(1) It should destroy the target and nothing but the target; it should

not leave a track of destruction or injury as the D. C. , Radio-frequen-

cy or Cryogenic probes do;   (2) The resultant lesions should be

(a) reproducible in size and predictable, (b) their size and shape

should be finely controllable, (c) they should be non-hemorrhagic,

(d) they should be discrete or sharply demarcated from surrounding

tissue which should be totally unaffected, and (e) they should be

instantaneous in development; (3) There should be no delayed effects

such as those produced by X-rays and other ionizing radiations; and

(4) With sub-lethal dosages it should be possible to produce transient

alterations of function to assist in target localization.

Focused ultrasound meets these requirements of an ideal surgical

tool [9:105]. It has the demonstrated ability to destroy preselected targets located deep within tissue without any damage to the tissue in the path or surrounding the lesion [10:484, 11:513, 12:502]. The margins of the necrotised tissue are sharply demarcated from the surrounding normal tissue [10:484, 11:513], the size of the lesion is predictable to a high degree of accuracy and can be finely controlled [11:513, 12:502], the lesions develop almost instantaneously [13,14] and there are no late effects such as those associated with ionizing radiations [10:484, 15:345]. Most of these apparently unique effects are attributable to the fact that many of the soft tissues are relatively permeable to ultrasonic energy and permit it to be focused deep within the tissues generating locally high intensities. Some other forms of energy such as infrared are almost totally absorbed at or near the surface. Thus, attempts to use a convergent cone of infra-red radiation [16] as well as a focused beam of coherent radiation from a ruby laser [17] for placement of lesions 15-20mm deep within the cat brain resulted only in superficial lesions. No deep lesions were produced. High frequency electromagnetic waves such as those used in medical diathermy and microwaves, on the other hand, do penetrate the tissues; but their wavelengths are too long for surgical purposes. The frequencies of ultrasound currently used are even lower; but as sound travels at much slower speeds than do the

electromagnetic waves, the wavelengths are very much shorter.
Focused ultrasound thus holds a unique position among physical
agents.

A simple collimated beam of ultrasonic radiation such as that
radiating from a plane disc transducer cannot produce trackless,
deep focal lesions, the formation of which depends on the creation of
a zone of high intensity at and restricted to the target. This can be
accomplished by either aiming two or more collimated or weakly
focused beams at the target from different directions [18:261], or by
the use of a single beam converging at the target [19:494]. The
former suffers from the inherent disadvantage that the energy content
in the side lobes is rather high and asymmetric; the latter is not
only simpler in principle, design and operation, but is probably also
more reliable and is being successfully used in several laboratories
of basic neural sciences without the continual assistance of physicists
and electronics technicians. At the Massachusetts General Hospital
it has also been used satisfactorily in patients for transdural irradiation of superficial cortical epileptogenic foci in cases of post-trau-
matic epilepsy, for performing commissural myelotomy of the spinal
cord for intractable pain [20], and for anterior horn neuronolysis in
a case of multiple sclerosis, in addition to the irradiation of a number

of cases of painful subcutaneous neuromata [21:858].

A general description of the technique will be followed by a discussion of problems particular to its surgical use. Clinical indications for placement of deep focal lesions are commonly to be found in the central nervous system (CNS), which is also amenable to stereotaxic procedures [22]. Though the brain will be used as an illustration, the following description is equally applicable to other soft tissues and organs provided that the acoustical properties of the target are not greatly different from those of the surrounding tissues (e.g., areas of calcification).

EQUIPMENT AND TECHNIQUE

Ultrasound is generated by the application of radio-frequency electrical power to a piezoelectric transducer. The transducer material can be either natural or cultured quartz or one of the ceramics, for example barium titanate, lead zirconium titanate, etc. Quartz has greater mechanical strength, lower internal friction, resists electrical abuse to a greater extent, and its function is less temperature dependent than that of ceramics. However, it needs comparatively much higher driving voltages. Ceramics are more efficient, need larger currents at lower voltages but tend to lose their piezoelectric

properties if mistreated electrically by overheating or by mechanical
shock. They are also inferior to quartz for operation at odd-multiple
harmonic frequencies. But because of the lower driving voltages
required, the complexity, size and cost of the generator are likely to
be lower. On these considerations, although quartz is the transducer
of choice for the development of ultrasonic techniques, the ceramics
probably are to be preferred for routine application.

The transducer may either be a section of a sphere or a flat disc.
In the former, the acoustic output converges at the centre, and the
focal length, the solid angle of radiation and the maximum working
distance are unalterable (Figure 1a).

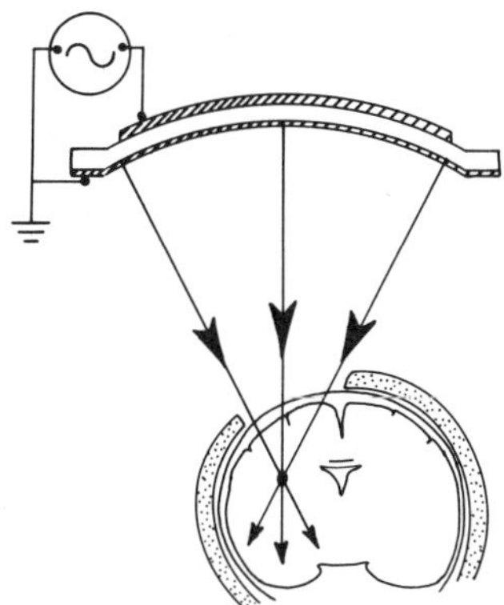

Figure 1a. Shows focusing with a bowl-shaped ceramic-transducer.
Because of crystallographic structure, quartz bowls are not satisfactory.

The output of the latter is a plane wave (a more or less collimated

beam) which is focused by suitable lenses (Figure 1b).

The lens is made of a plastic, such as polystyrene or rexolite, is

of simple design and is fabricated easily and inexpensively. A disc

transducer-lens combination offers flexibility in the choice of focal

lengths and thus the solid angle of radiation, beam geometry (spheri-

cal, cylindrical or compound focusing) and the shape of the focal

region (needle shaped or almost spherical [23]). The lens is coupled

to the transducer by a thin film of silicone oil and is held concentric

with it by a retainer ring (Figure 2). With the irradiation head

machined to the tolerances common in the industry the focus of the

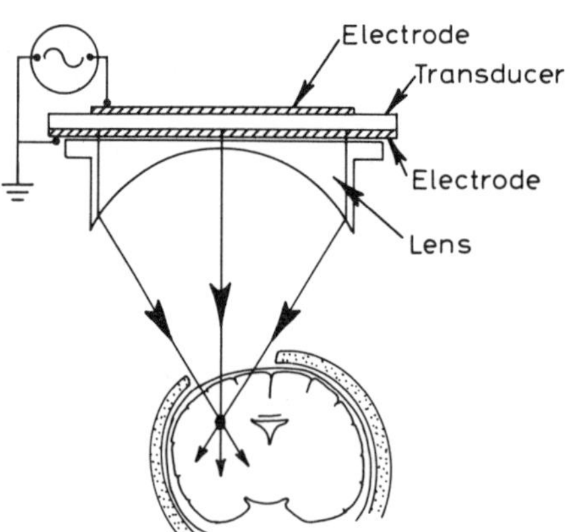

Figure 1b. Shows disc-transducer/lens combination. Disc may be
ceramic or quartz.

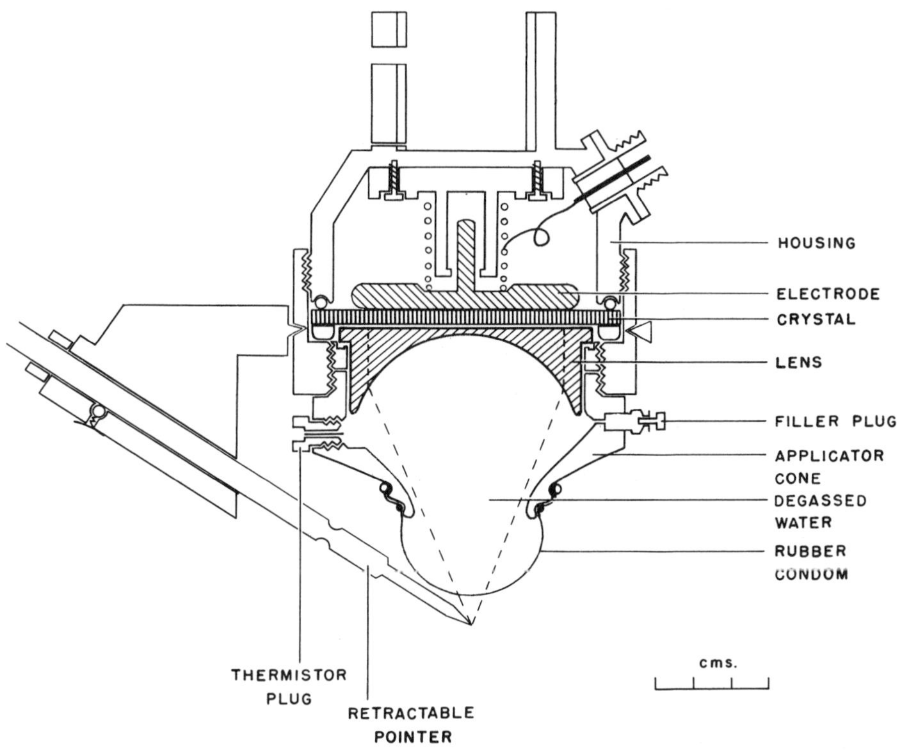

Figure 2.   Diagram of assembled irradiation head.   See text.

system, in relation to the radiating surface of the transducer, is as

constant for any given focal length as it is for a focused bowl - even

though the irradiation head may be dismantled occasionally for the

use of a different lens or for renewal of the silicone oil coupling.   Up

to 40 percent of the acoustic energy emanating from the transducer

may be lost in the lens, depending on its material and average thick-

ness.   But the focused output is still more than adequate to meet all

probable power requirements.   Moreover, it will be seen later that

the use of excessively high power levels should be avoided as it is
related to the occurrence of hemorrhages. The intensity distribution
(and thus the focal length) at the focus in water is determined experimentally using a fine thermocouple embedded in soft polyethylene [24].
Relative intensity distribution in the two axes for a spherical and an
elliptical lens is shown in Figure 3.

The size of the smallest lesion that can be made is inversely related to the frequency of the ultrasound, and thus higher frequencies
would appear to be more suitable. For the placement of deep lesions
at high frequencies much more power is needed at the transducer than

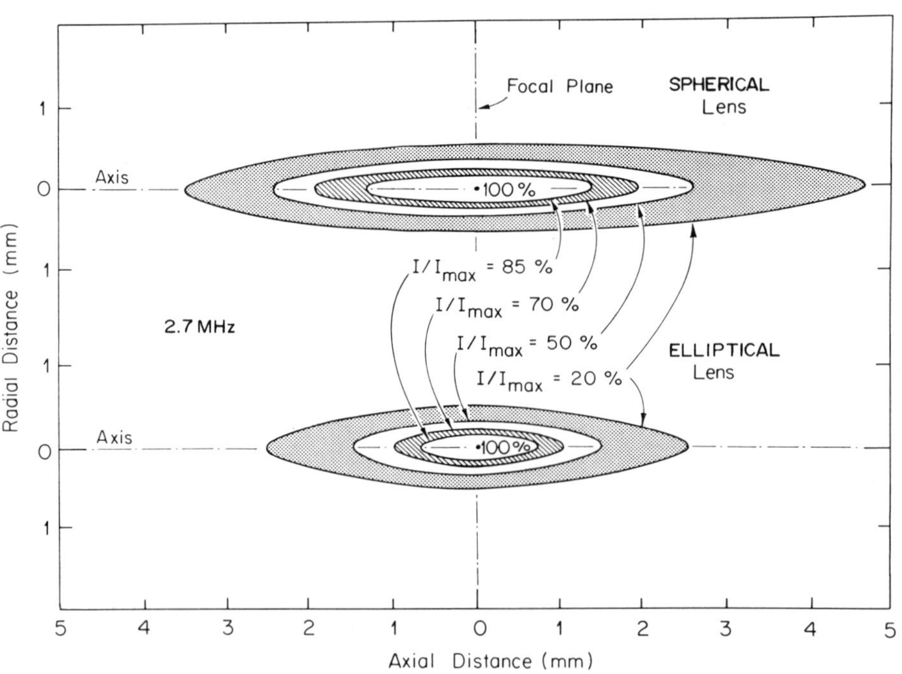

Figure 3. Map of lines of constant intensity at the focus, showing the
relative intensity distribution in the two axes.

at lower frequencies since the attenuation in the intensity of sound in
its passage through intervening soft tissues is directly related to the
frequency. And since very small lesions deep within the brain are
clinically not manifest, frequencies 1, 3 and 5 MHz cover most of the
actual needs. The lower frequencies are used for deep lesions in
large structures and the higher for lesion placement in small structures such as the spinal cord, where the lesion size is critical. A
transducer ground finely to a fundamental resonant frequency of 1 MHz
in the thickness mode is used to generate all the three frequencies.

The convergent beam of ultrasound emanating from the lens is
transmitted to the soft tissues (the exposed dura mater of the brain)
through degassed water contained in the applicator cone (Figure 2).
Since the acoustic impedances of air and of bone are very much
different from that of water and soft tissues even the smallest bubble
of air or spicule of bone lying in the path of sound reflects the energy
and distorts the beam geometry, (Figure 4). Meticulous care is
therefore necessary to prevent either air or bone from protruding
into the cone of sound. When subjected to an intense ultrasonic field
fluids release dissolved gases in the form of small bubbles. To
avoid this, water and saline solutions are degassed thoroughly by
boiling for about an hour prior to their use in ultrasonic procedures.
After cooling to 37°C under partial vacuum they are stored in air-
tight, rubber hot water bottles and maintained at a temperature of
37°C.

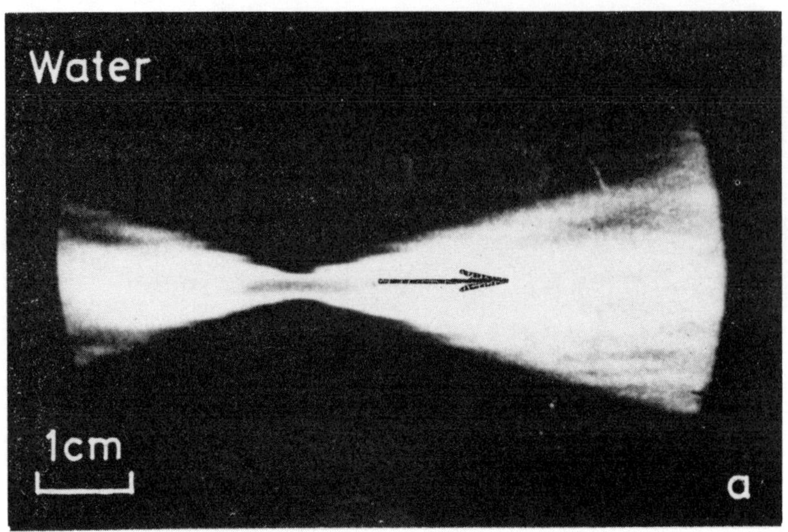

Figure 4a. Schlieren photograph. The arrow indicates the direction of the main axis of radiation. Transducer was 8 cm to the left of the focus. The generator was slightly detuned to show the shape of the high intensity focus, which in the photograph is 'burnt in'.

The focal length of the system with any particular lens is temperature dependent. To maintain a constant focal length, in spite of changing target depths and varying ambient temperature, the temperature of the components in the ultrasonic pathway is maintained at 37°C - the average temperature of mammalian tissues <u>in vivo</u>. A graded series of applicator cones with different working distances as well as a telescoping cone [25] permit quick and easy coupling with different target depths. The sac contacts the dura through a thin film of saline solution and conforms to the shape of the brain. There is thus little chance of air bubbles being trapped in the path of the sound. The hydrostatic pressure in the cone and the sac is normally

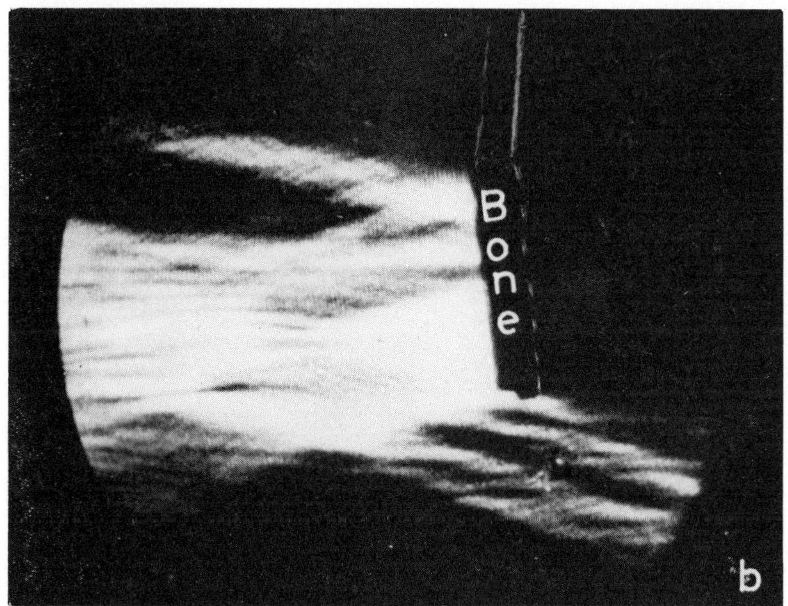

Figure 4b. Schlieren photograph. The arrow indicates the direction of the main axis of radiation. Transducer was 8 cm to the left of the focus. A preserved specimen of bone was placed in the field distal to the focus. Note the reflections, the distortion of the beam geometry and the standing waves, which are probably exaggerated, due to the tissue elements in the cancellous part of the bone having been replaced by air during its processing for preservation.

atmospheric but, if necessary, can be raised to prevent pulsation of the exposed brain. The rubber is thin and does not reduce the sound intensity measurably unless it lies in the focal plane - a situation that cannot arise in placement of deep lesions.

Presence of bone in the ultrasonic path not only distorts the field by reflection, but may also destroy the underlying tissues in contact with it by absorbing ultrasonic energy and dissipating it as heat. The skull also poses some difficult problems in ultrasonic diagnosis of intracranial pathology, as we shall see later. A craniotomy, adequate in extent to permit unimpeded passage of the cone of sound, is imperative. The extent of the craniotomy depends on the solid angle of radiation and the depth of the target from the cranial surface. The larger the angle and the deeper the target the larger the size of the craniotomy needed (Figure 1). An optical projector, interchangeable with the irradiation head, projects a cone of high intensity cold light of the same dimensions as the cone of ultrasound and is of considerable assistance in determining the extent of needed craniotomy on curved or irregular surfaces [26]. A further check on the adequacy of craniotomy and of the coupling is provided by the pulse-echo lesion detection system discussed later.

For stereotaxic procedures the irradiation head is mounted on a X-Y-Z positioner carrying the stereotaxic head holder so that each of the coordinates reads zero when the ultrasonic focus is exactly at the Horsley-Clark zero. To assist in irradiation of structures not amenable to stereotaxis, such as the spinal cord where the coordinates of the target have to be determined with reference to the posterior median sulcus, a rotating retractable pointer is set to indicate the focal length.

Radiation pressure, particle velocity and intensity of ultrasound are interrelated and Heuter [27] has shown that any of the three may be used to describe adequately the ultrasonic conditions of the focal point of a single focusing transducer. Radiation pressure measure-

ments can be performed easily and rapidly by radiation pressure
gauges calibrated to read the output in watts directly [28]. The
measurements can be converted into units of particle velocity or intensity for comparison with results of other workers. In practice,
by actually setting the ultrasonic output to the level desired, it is
possible to reproduce the irradiation conditions accurately. Lesion
size is found to be related to radiation pressure in a reproducible and
consistent manner.

FUNCTIONAL TESTS

To test rapidly the proper functioning of the equipment, and to establish the repetitive stability of the acoustical output, the results of
irradiation of a stable and relatively homogeneous medium such as
methacrylate (plexiglas, lucite, Perspex) are analyzed [29:412].
Transparent, clear and stainfree bars of methacrylate, 18mm square
and about 45 cm long, buffed and polished on all surfaces, are used.
Irradiation of such a bar with an adequate dosage of ultrasound and
with the focus placed beneath the surface results in the development
of stable, trackless, discrete, egg-shaped areas of stress within the
bar, readily visible in polarized light. For the same irradiation
parameters, the dimensions of these 'lesions' in the three planes are
remarkably constant. There is a consistent and reproducible relationship between the ultrasonic dosage and the size of these lesions
(Figure 5), comparable to that obtained in the brain of the experi-

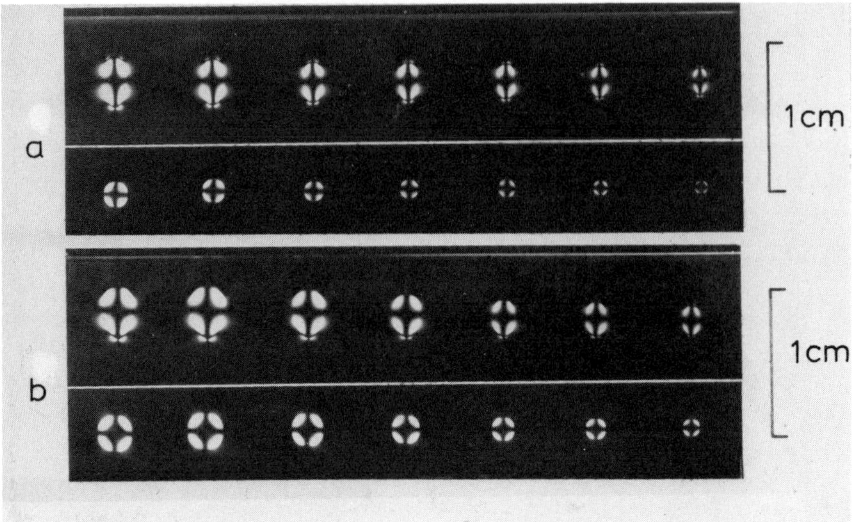

Figure 5.   Photographs of two sets of lesions in methacrylate bars.
In each set, the view at right angles to the direction of radiation is
seen at the top, and the view in the axis of radiation, at the bottom.
(a) Lesions were made with single pulses.   The pulse duration was
decreased in steps from left to right.   (b) Lesions were made with
multiple pulses of constant pulse duration.   The number of pulses
used decreased from left to right.   Note the greater sphericity
(diameter/length) of lesions in (b).

mental animals.   The use of methacrylate as a preliminary test

medium permits rapid evaluation of varying ultrasonic dosage param-

eters as well as the repeatability and long term stability of the ultra-

sonic output.   Tests of ultrasonic generators (with different harmonic

content in their output, for instance) show that the plastic can indeed

be used as 'phantom brain' in screening of ultrasonic equipment.

RESULTS IN ANIMALS

Over the last 10 years the brains or spinal cords of more than 1,600
animals have been subjected to over 10,000 irradiations by the author
and his colleagues, the total number of lesions placed being well over
7,500. The time required for each operative procedure, involving
stereotaxis, in the cat or monkey, from the administration of intra-
venous anaesthetic to wound closure is about 20 minutes of which
ultrasonic procedures account for approximately 3 to 5 minutes; the
irradiation itself lasting 0.5 to 2.0 seconds. The set-up time for
ultrasonic procedures, namely filling and closure of the applicator
cone, temperature stabilization and radiation pressure calibration
take about 10 minutes. Many of the irradiation procedures have been
carried out by a nurse-technician, without assistance except during
the administration of anaesthesia. For non-stereotaxic procedures -
experimental or therapeutic - the amount of time required increases
sharply - most of it being consumed by preoperative positioning of
the target organ and its immobilization for the alignment of the focus
with the target and irradiation. The postoperative mortality and
morbidity are low, which is attributed to the fact that the irradiation
is transdural and thus the chances of infection are reduced. The
intracranial fluid dynamics are also relatively undisturbed since

there is no less of the cerebro-spinal fluid.

In an extensive study of the effects of irradiation on the CNS of experimental animals [11:513], with suprathreshold irradiation dosages, a trackless, pan-necrotic, focal lesion restricted to the irradiation target (Figure 6) was evident in every single instance. The outstanding feature of the results obtained was the reproducibility of the effects of irradiation at each of the dosage levels, with single or multiple pulses, the coefficient of variation being about 10 percent. The lesion dimensions were directly related to dosage levels in a statistically significant manner (Figure 7). Smallest lesions are almost spherical, larger lesions are progressively more elongated (Figure 6) corresponding to the shape of progressively lower iso-intensity-lines (Figure 3). At a frequency of 2.7 MHz the smallest lesion that could be made with certainty was approximately 1.2 mm in length and 0.4 mm in diameter. The length of the largest lesions that could be consistently made was 11 mm with single pulses and 17 mm with multiple pulses. Like those in the plastic bar the lesions were egg-shaped, the sphericity (diam. length) of those from multiple pulses being greater than that of single pulse lesions (Figure 8). For maximum sphericity the optimum interpulse interval was about 1.0 sec. Sphericity was also related to the solid angle of radiation;

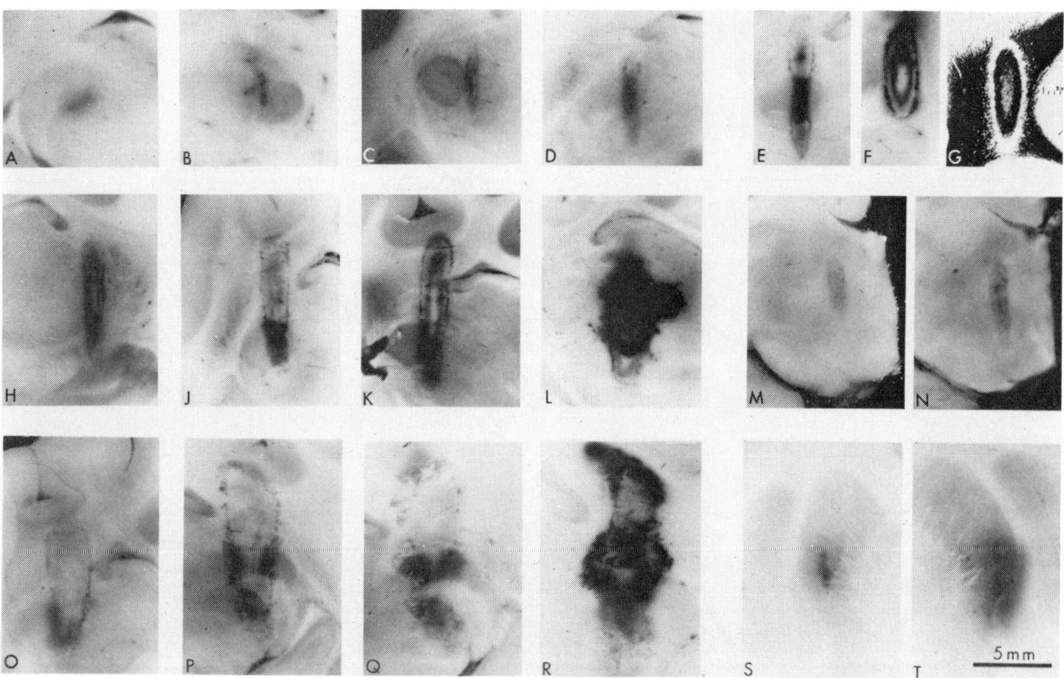

Figure 6.   A, B, C, D, H, J, K, and L are photographs of lesions produced in cat brain with a single pulse. Note diffuse blueing in A and
hemorrhage in L.   E shows a typical constriction in diameter when
lesion produced with a single pulse passes through grey matter with
normal circulation; F a lesion in subcortical white matter in monkey.
G shows a section of a lesion in subcortical white matter in cat 72
hours after irradiation.   Loyez myelin stain; note different scale
(0. 5 mm). O, P, Q and R show lesions produced by multiple pulses.
Note greater diameter compared to H, J, K, and L and hemorrhage in
R.   M, N, S, and T show influence of circulation on lesion size in grey
matter.   M and S are lesions from irradiation with normal circulation; N and T are from irradiation during cranial circulatory arrest.
All lesions except G were stained with Trypan Blue and photographed
at magnification indicated in T.

the smaller the angle, the less spherical the lesion.

All lesions were found to be sharply demarcated, discrete, friable

areas of coagulative change.   No evidence of any hemorrhage was

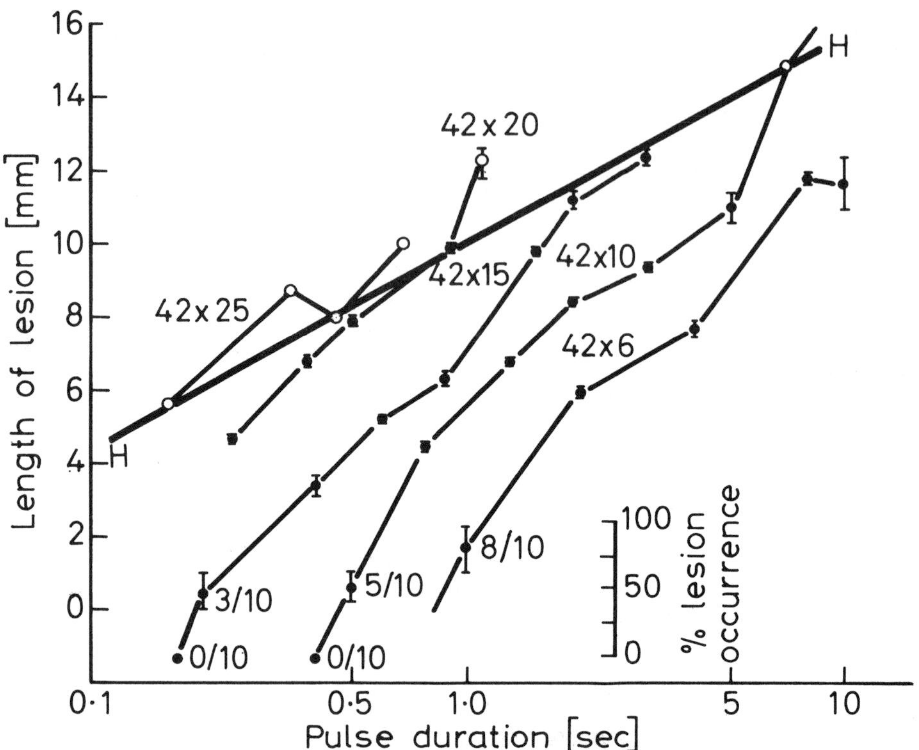

Figure 7. Relation between pulse duration (log. scale) and the length of lesions with a single pulse at average focal intensities of (42 X) 6, 10, 15, 20 and 25 W/cm^2 respectively. Each point indicates mean of ten lesions and the standard deviation. Probability of lesion occurrence where not shown in 100 percent. O, hemorrhage distending the lesion and spreading intracerebrally when severe. Line H-H shows the safe limit below which hemorrhage was never encountered.

even encountered in lesions of up to the maximum size referred to above (Figures 6 and 7). But all attempts to increase the size of the lesion - particularly by the use of a single pulse of high intensity - invariably produced hemorrhages distending the lesion, spreading intracerebrally and often rupturing into the lateral ventricles.

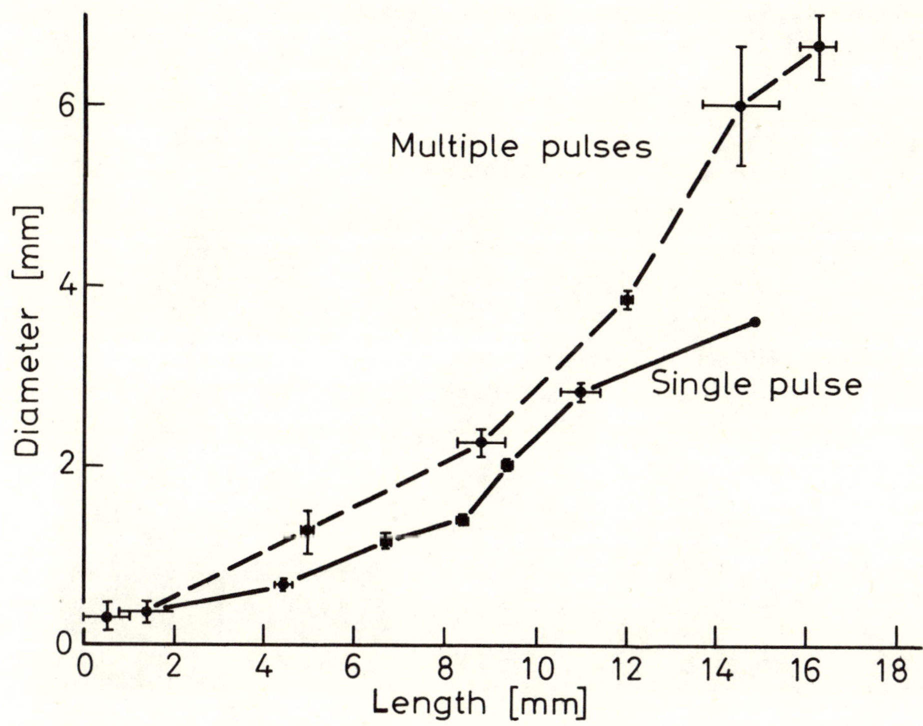

Figure 8.   Relation between length and diameter of lesions made with single (—) and multiple (- - -) pulses.

The volume of the tissue necrotised can be increased safely by juxta-position (with slight overlapping) of two or more individual lesions.

No damage was detected in the brain in the path of the sound nor in the periosteum or bone underneath the deepest lesion (Figure 9, P).

FACTORS AFFECTING LESION SIZE

The attenuation sound in the 'normal' dura mater of the cat is about

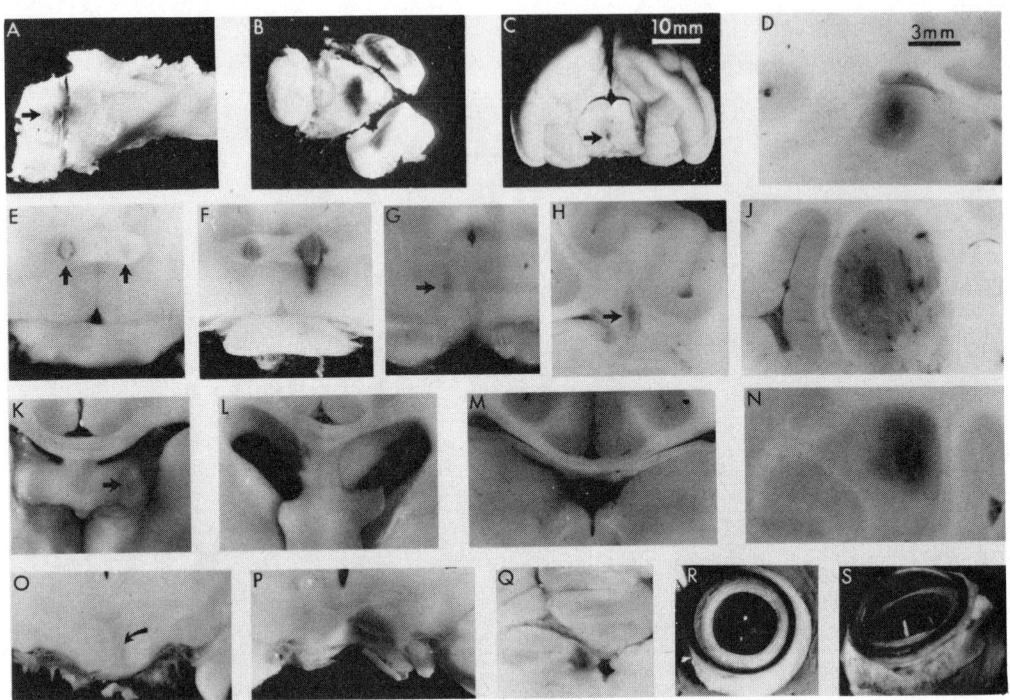

Figure 9.   Photographs showing lesions in: A, maxillary nerve at
Gasserian ganglion; B, pituitary body; C, G, Edinger-Westphal
nucleus; E, F, anterior commissure; H, subcortical white matter;
K, fornix; O, P, mammillary body; Q, XII nerve nucleus; R. S, lens
oculi; D, J, N, grey matter near grey-white interface; L, M, show
extremes of variability encountered in lateral ventricles at F + 13. 0.
All photographs except C, R, and S are at magnification indicated in
D.

5 percent.   Presence of dural adhesions, as well as of scars, was

found to be one of the sources of variation in the results of irradia-

tion.   In patients with intracranial pathology this is likely to be a

bothersome — though a minor — problem unless the dural attenuation

can be determined without opening it.

The target depth can influence the dosage-lesion relationship by attenuation and refraction. However, both of these can easily be estimated for and corrected. It must be remembered that attenuation is frequency dependent whereas refraction is velocity and temperature related. In the cat the presence or absence of the superior sagittal sinus and the lateral ventricles in the path of ultrasound was not found to influence the dosage-lesion relationship. However, the presence of air or other radio-opague material within the ventricles might have profound influence on the ultrasonic beam geometry.

With the same ultrasonic dosage, the lesions in grey matter were found to be significantly smaller than those in the white matter. Lesions in grey matter, surrounded by white matter, showed constriction in the region of grey matter (Figure 6E). This exemplifies the common statement that the threshold of the white matter is lower than that of the grey matter. Figure 10a shows a section of à cat brain with a planar lesion - the grey matter of the cortex is intact and the underlying white matter is extensively destroyed. When the spinal cord is irradiated similarly it is found that it is the grey matter that is completely destroyed leaving the surrounding white matter almost intact (Figure 10b). Thus the thresholds appear to be reversed. Work in progress indicates that the direction of the fibres

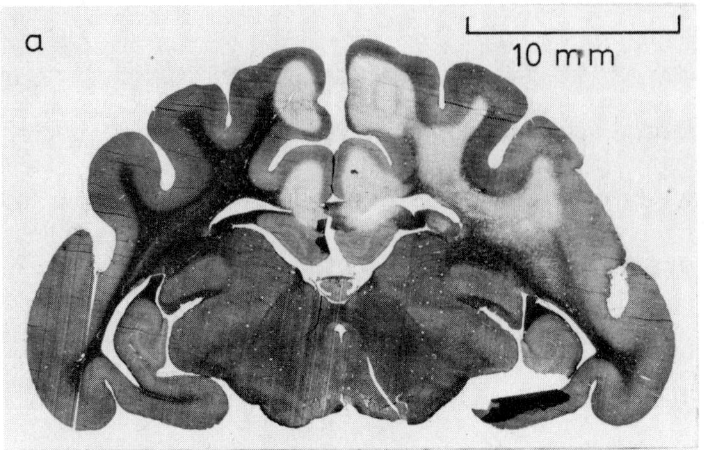

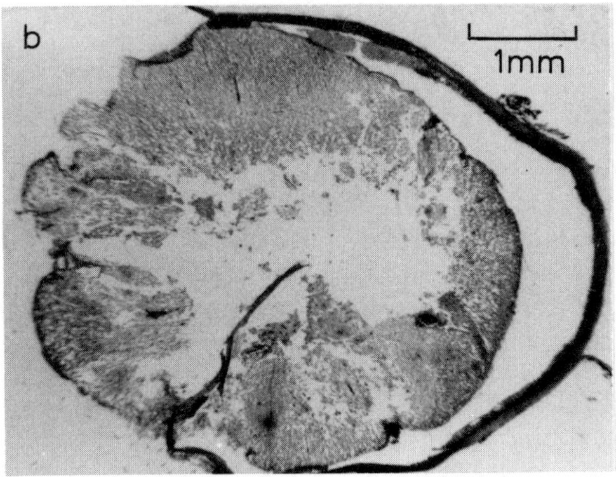

Figure 10.   Planar lesions.   (a) Brain of the cat. (b) Spinal cord of
the cat.   The survival period in each was one week. Loyez myelin
stain.   Most of the grey matter in (b) was friable and has dropped off.
The dura mater was intact.

in the white matter, relative to the direction of the main radiation

axis, is an important factor governing the apparent sensitivity of the

white matter to ultrasonic destruction.   In the CNS the intrathecal

parts of the spinal nerve roots appear to be the elements most resistant to ultrasonic destruction.

The state of local circulation also affects the lesion formation. Arrest of cranial circulation during irradiation with near threshold dosages resulted in larger lesions both in grey and white matter. At suprathreshold dosage levels, the effect was smaller and restricted to grey matter alone; the blood perfusion through which is much higher than that in the white matter. The lesions then tended to approximate to those in white matter in size [11:513]. The constriction in the grey matter, referred to above, disappears in lesions made with cranial circulation arrested. Examination of control specimens by routine histological methods did not reveal any damage attributable to anoxia.

The temperature of the tissue at the time of irradiation has a profound effect on the dosage-lesion relationship. The lesion size is directly related to the temperature (Table 1). In addition to the evident implications of this observation to the mechanism of lesion formation it is of considerable practical importance in surgery since so many of the surgical procedures are now carried out under hypothermia.

Fry and Myers [30:315] have reported irradiation through intact scalp after previous craniectomy, the advantages of which are ob-

Table 1. Relationship of ultrasonic lesion size in brain to temperature of the tissue: cat and dog. Average focal intensity= $42 \times 10 W/cm^2$; pulse duration = 1.3 sec., single pulse.

Core temperature [^{o}C]	Number of observations	Lesion Volume [mm^3]	Length [mm]	Diameter [mm]	Sphericity d/1
37	5	4.94	6.74	1.18	0.175
31	5	1.25	4.36	0.74	0.170
24	5	0.14	1.23	0.41	0.333
22	10	No lesions produced			

vious. However, data on the ultrasonic properties of human scalp are not readily available. The scalp of furry animals - lacking sweat glands - apparently is not comparable to the human scalp. Scalp of the adult cat was found to attenuate as much as 66 percent of the ultrasonic energy. Lesions placed deep within the cat brain by irradiation through the scalp were found to be variable in size and, in each instance, the scalp showed signs of damage. The scalp and the skull of the foetus and the newborn, on the other hand, have acoustic properties similar to other soft tissues and lesions of controlled size could consistently be placed within the brain of foetal rabbits by irradiation through intact scalp and skull [12:502].

It is evident that any factors which alter the acoustic properties or the thermal characteristics of the tissue in the path of ultrasound would influence the dosage-lesion relationship. Currently the ultrasonic dosage is determined empirically whether it be in the plastic bar or the brain. Temperature gradients in the focal region in plexiglas have been studied by incorporation of thermochromic materials [31] and by embedded thermocouples [32]. The resulting heat calculations indicate the lesions in plexiglas are probably thermal in origin. A study was therefore undertaken to determine if thermal mechanisms alone can explain the development of the lesions and all of their measurable characteristics in plastic as well as in brain. A purely thermal model was assumed and analytical prediction of lesion development and lesion size and shape for varying values of ultrasonic and thermal constants and controllable variables (frequency, focusing, dosage, target depth, etc.) was attempted [32]. An empirical equation to describe the axial and radial ultrasonic energy distribution at the focus in water was derived. Appropriate heat transfer equations were developed for temperature distribution resulting from ultrasonic irradiation. The computed temperature profiles were plotted against non-dimensionalized parameters. Temperatures at the lesion center and the lesion boundary were determined experimentally and were found to rise up to 400°C and 150°C in plexiglas and 100°C and 58°C in brain (Figure 11) respectively. Expected

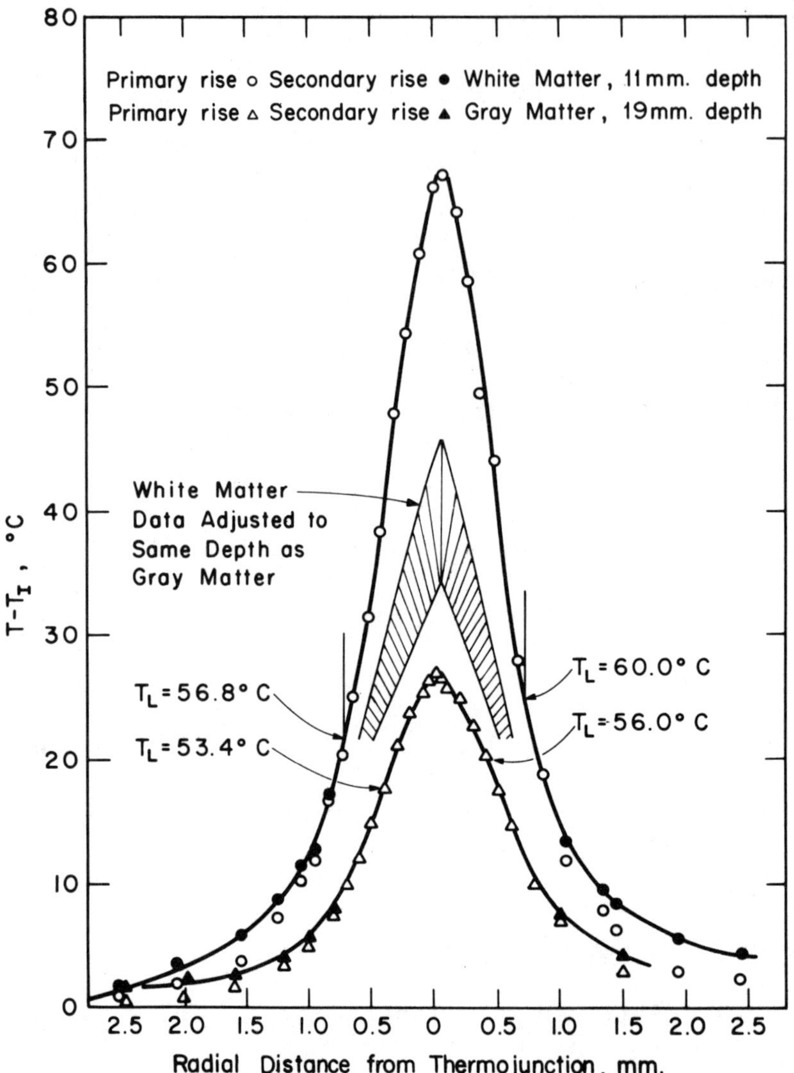

Figure 11.   The temperature profile measured in cat brain in vivo. Frequency 2.7 MHz; pulse duration 2.0 sec.; acoustic power 10 watts; chromel-constantan thermocouple 0.002-inch diameter; maximum temperature was reached at the end of irradiation (primary rise) with temperature rises of 20° C or more. T_L - lesion boundary temperature.

axial and radial lesion dimensions were read off the curves of tem-

perature distribution at the measured lesion boundary temperature.
Comparison of these predicted lesion dimensions with experimental
data (Figure 12) indicates that lesion development in the brain (as in
the plastic) at the frequency and power levels used in these studies
can be explained by purely thermal considerations. The range of
field variables (frequencies and power levels) as well as state variables (base temperature of the tissue, circulation, type of tissue and
consideration of anisotropy, etc.) over which the above observations
of a relatively constant lesion boundary temperature is valid is
currently being studied. It is expected <u>a priori</u> that the duration over
which the temperature rise is sustained will influence to some degree
the extent of damage to the tissue. These ultrasonic experiments are
being repeated using purely thermal sources for elevating the tissue
temperatures. Preliminary results indicate that the extent of damage
can be best correlated with the magnitude of temperature rise and its
duration whether produced ultrasonically or thermally.

The irradiation system described here permits accurate lesion
placement in any preselected target of white or grey matter, regardless of its location within the CNS (Fig. 9). In a large series of
blind experiments, using six different target sites, it was found that
the probability of lesion placement is restricted only by the variability in the stereotaxic location of the structure itself [11:513].

The histopathological features (Figure 13) of the ultrasonic lesion
in adults and foetuses have been studied starting immediately after

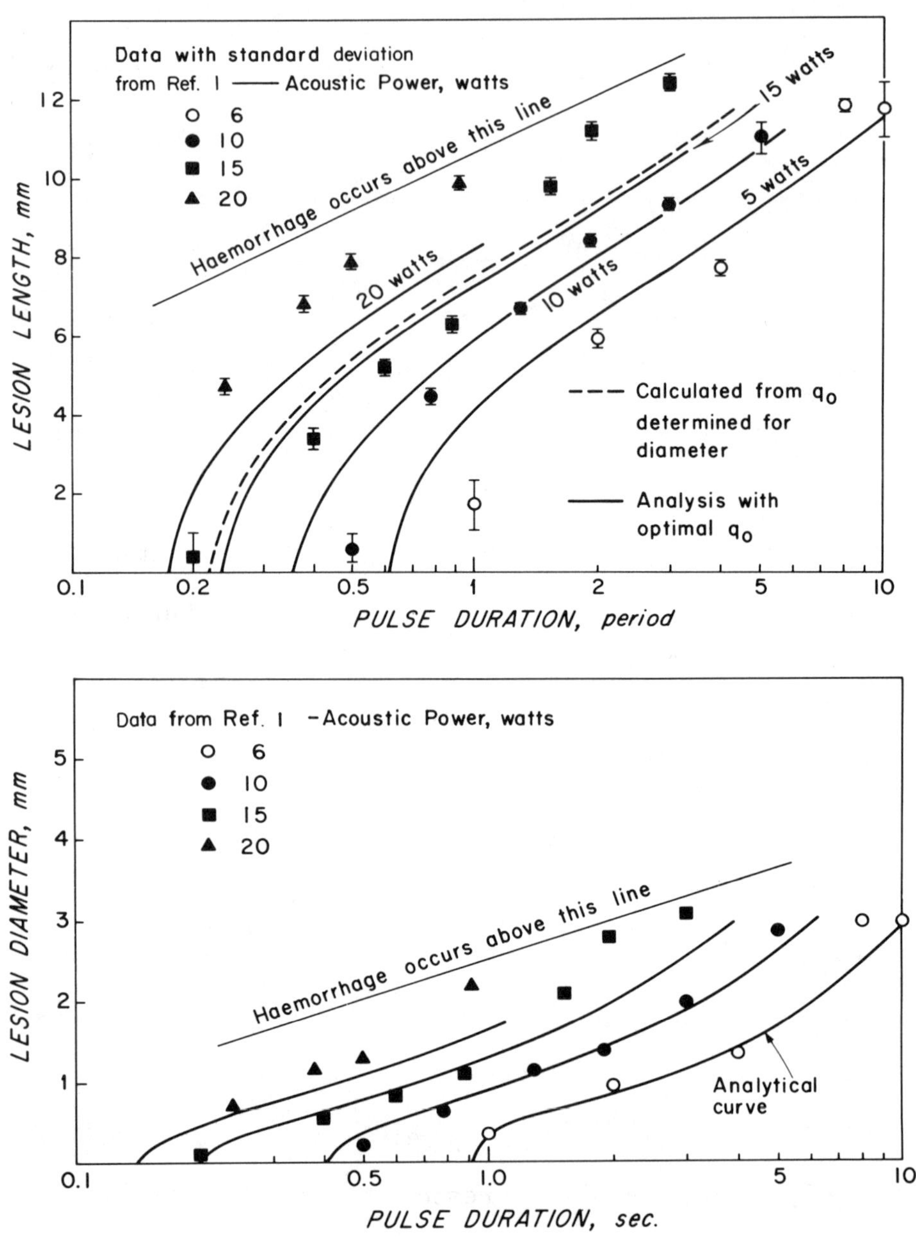

Figure 12.   A comparison of experimental data and analytical results in cylindrical coordinates for lesion diameter in cat brain at various pulse durations. $q_0 = 338$ Watts/cm^3 at 10 Watts acoustic power, empirical value.

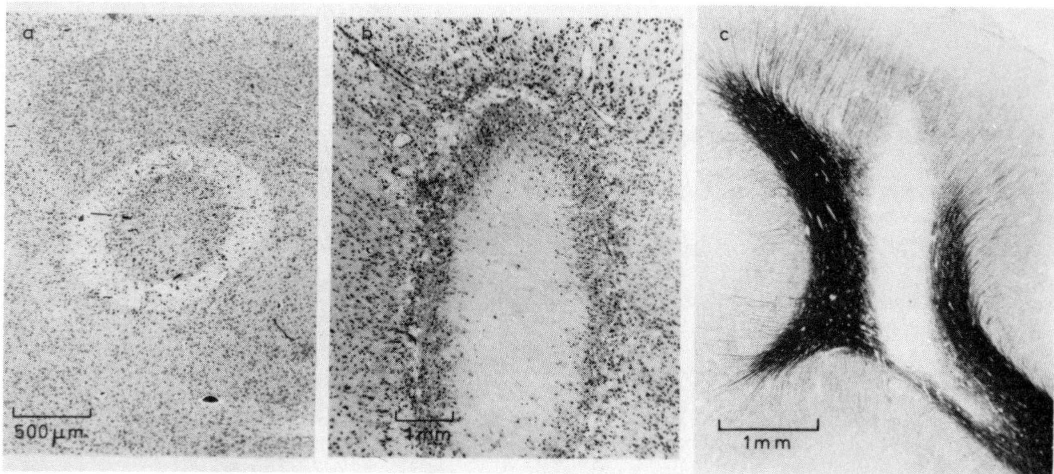

Figure 13.   (a) Photograph of a lesion 20 mins. after irradiation of
a 2-week old rabbit.   Note sharp border and the characteristic moat-
island form. H.E. stain.   (b) Lesion 6 days after irradiation in cat.
Note that the central core is solid and is resistant to lysis, though
only very few cell remnants are visible.   The peripheral zone is filled
with macrophages, microglia and contains many blood vessels.
Cresyl violet stain.   (c) Lesion 6 months after irradiation in cat.
Note the sharp borders of the scar tissue and the Wallerian degeneration.   Loyez myelin stain.

irradiation and afterwards at intervals up to 2 years [9:105, 10:484,

12:502].   The lesions are pan-necrotic, all tissue elements including

the blood vessels being completely destroyed within the lesion. They

are sharply demarcated from the surrounding healthy tissue. All but

the smallest lesions have a characteristic moat-island form consist-

ing of a coagulated central core resistant to lysis and a margin of

liquefaction.   Healing occurs by phagocytosis of necrotic tissue by

microglia and macrophages and the concurrent production of astro-

cytic gliosis and, in about 4 months, the lesion is replaced by a glial scar. There is no delayed local or remote spread of the lesion except by Wallerian degeneration. The ultrasonic lesion and the scar are non-irritative and are electro-encephalographically 'silent'. The earliest change is difficult to see except as a reduction in the staining ability of the tissue [13]. But with E. M. (Figure 14) it is restricted to the mitochondria which appear swollen and lack electron density [14]. This only indicates that the tissue has undergone some change which may or may not be reversible and can also be seen in muscle fibres after exercise. In later specimens at the same dosage

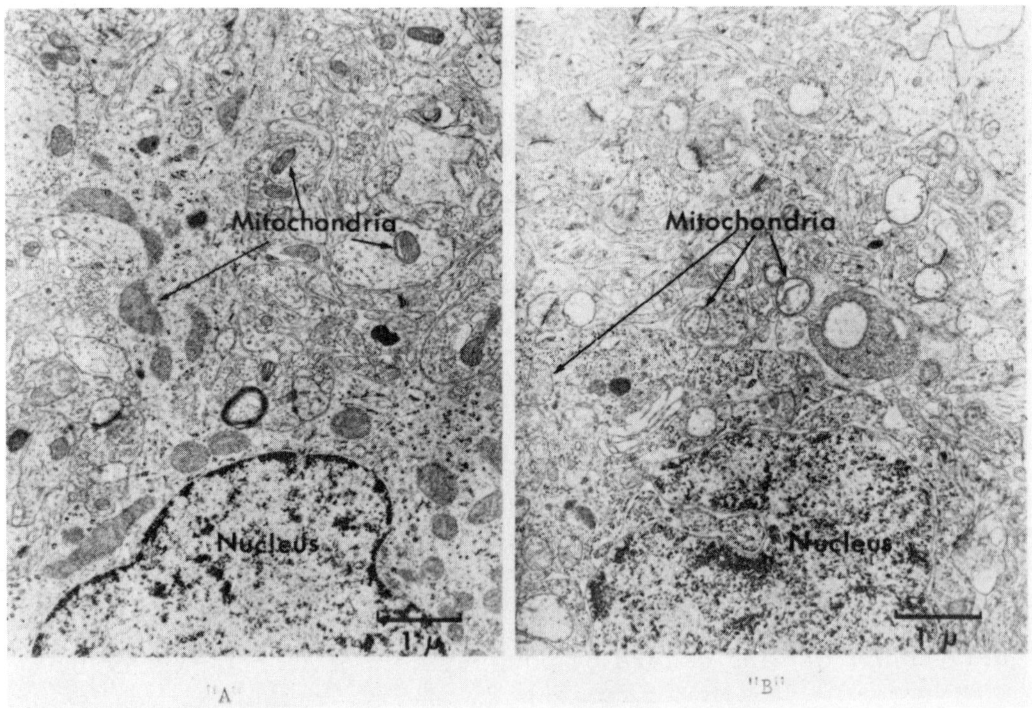

Figure 14. Electron micrographs, cat brain. (A) Normal cortex (grey matter). (B) Cortex immediately (approx. 1.5 min.) after ultrasonic irradiation.

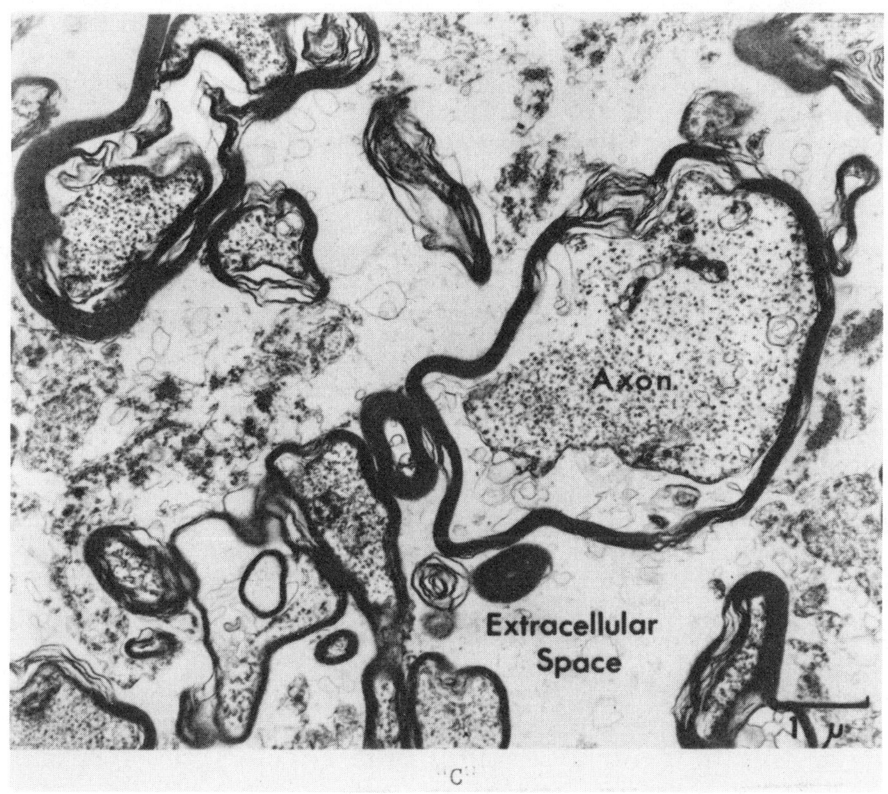

Figure 14.   Electron micrographs, cat brain. (C) Subcortical white matter 2.5 hrs. after ultrasonic irradiation. Note in B the swelling of the mitochondria and the deformation or disappearance of their cristae. The electron density in the matrix of the mitochondria is markedly diminished and some appear as vacuoles surrounded by double limiting membranes. Note that there is remarkably little change in the remaining ultrastructure and no evidence of mechanical disruption of structural organization. In C the axoplasm of the axons appears as though coagulated to form discrete particles about 500 to 1000 Å in diameter. The dissociation of myelin lamellae may be a fixation artifact. The extracellular space is remarkably expanded and filled with material of low electron density. Changes comparable to those in B and C can also be produced by heating a wire embedded in the brain or by infrared irradiation.

level, however, large pores are seen in the plasma membranes

indicating the presence of irreversible damage leading to the death of the cell. In the early specimen the structural organization of the

tissue shows no signs of disruption such as could be attributed to

any mechanical effects of ultrasonic irradiation. Ultrasonic changes,

comparable in every respect to these are also produced by thermal

irradiation.

SURGICAL APPLICATIONS

With all these virtues, it is not surprising that in basic neural sci-

ences (and for non-stereotaxic superficial lesion placement in clini-

cal practice) single beam focused ultrasonic radiation is the tool of

choice for the researcher (or clinician) who needs precise lesion

placement but no tracks to confuse the neuronantomic or neurophys-

iological observations--and who is prepared to make a trephine hole

instead of a burr hole in the skull. And it is this factor, the size of

the craniotomy, that has prevented the routine use of the technique in

surgery of deep structures. In man many of the targets of neuro-

surgical importance lie 6-10cm from the cranial surface and cran-

iotomies of up to 10cm were required. But now, with improvements

in the focusing, the solid angle of radiation can be reduced to 35° and

still produce lesions of acceptable shape. This reduces the size of

craniotomy needed to a more practicable diameter of up to 6.5cm.

In man craniotomy, of even this reduced size, leads to a shifting of

the brain from its normal position which adds to the difficulties in

accurate anatomical localization of the target areas. Even without craniotomy, stereotaxic location of brain structures, that is their

location relative to external cranial landmarks, is not as precise in

man as it is in the cat. Thus, injection of some radio-opaque mater-

ial into the ventricular system is necessary to determine the exact

location of internal landmarks with respect to the skull. Since the

acoustic impedance of radio-opaque materials (air to iodine com-

pounds) is very different from that of the brain, ultrasonic irradiation

has to be delayed until they are completely excreted--with the possib-

ility that the brain structures may shift again. Fry and Fry [33:171]

have reported on the use of a water soluble iodine compound (Conray-

Mallinckrodt Chemical Works) which is apparently excreted within

10 mins. Use of this compound in suitable dilutions may improve

this situation considerably.

In radio-frequency or cryogenic lesion techniques the electrode is

inserted towards the target under radiological control. The same or

another adjacent electrode is then used to record the electrical activ-

ity (resting or evoked potentials) from the tissues at or near the ad-

vancing tip, until the correct response is obtained. If a satisfactory

response is not obtained in one track the electrode is reinserted a

little distance away and the tip advanced gradually. Each track is

small but necessary, and after a number of insertions, the total

damage may not be inconsiderable, though it is not clinically mani-

fest, at least with the current state of knowledge of the function of the

CNS and the techniques of examination. The ultrasonic beam cannot be inserted under radiographic control. Unless the trackless beam itself can be used for anatomical and/or functional localization of the target, the raison d'etre of the technique is defeated and it cannot-- in spite of all its other virtues--compete with the current techniques for placement of deep focal lesions stereotaxically. Hence the importance of the so-called 'reversible lesions'; these are ultrasonically-induced temporary functional changes, for which no structural changes can be demonstrated, at least by the time when the function is fully restored. Only one single instance of this phenomenon in man is on record, supported by studies on peripheral nerves [34:47] in the visual system of the cat [35:281, 21:858] and the sympathetic ganglia [36]. The need for extensive and critical statistical studies aimed at confirming the occurrence of reversible functional changes in each of the clinically important target systems and establishment of the dosimetry cannot be overemphasized. The problem is complicated by the requirement of certainty of the absence of any lesion. But, presumably, a lesion which escapes careful examination of serially cut sections by a competent neuropathologist, would not be clinically disastrous--certainly no more so than the track of an electrode. In a current study the approach has been to develop a technique of detecting the lesion as it is formed by incorporation of a pulse-echo system connected to the same transducer that is used for irradiation [37:451, 38]. The detection level of the system can be set to signal incipient lesion formation. This permits rejection of the results in instances where there might be even a remote chance of lesion existence and to concentrate on only those results and specimens where the irradiation is known to have been at subthreshold dosages. The system is also useful in confirming the adequacy of the craniotomy and of the coupling and provides, in addition, a rather simple and accurate method for rapid determination of the focal length of the system.

The answer to the question of reversible functional effects is crucial to the future of focused ultrasonic radiation in surgery and unless it is answered satisfactorily the efforts to marry the rather heavy irradiation head to a stereotaxic apparatus would appear to be premature and possibly futile. In experimental sciences focused ultrasonic irradiation is the method of choice for placement of focal lesions and is being successfully used in studies of the connections of otherwise inaccessible brain structures such as the anterior commissure [39] and the insula [40]. A confirmation of the existence of the more exotic effects, such as size-- or component--differential lesions, would only enhance its usefulness. But so far as the surgery of deep structures is concerned, its status is still _sub judice._ The reason for this, of course, is the presence of the skull.

Now, going back to the diagnostic applications, you may recall that of all the body regions, I did not mention two--viz the head and the chest. In the head, the existing ultrasonic systems for detection and localization of intracranial pathology, e.g. hematomas, aneurisms, tumors, etc. are severely restricted in accuracy and reliability because of the unpredictable effects of the skull on the acoustic transmission. These effects are due to the variable thickness and irregular shape of the skull, the differences of sound speed in skull (4080m. sec^{-1}) and brain (1540m. sec^{-1}), the relatively large and variable absorption ($\alpha/f = 20dBcm^{-1}MHz^{-1}$) in the skull compared to that in the brain ($\alpha/f = 0.85dBcm^{-1}MHz^{-1}$), and the inhomogeneous structure of the bone. Attempts are being made to compensate for the skull effects on the echoencephalograms by simultaneous measurement of local skull characteristics [41]. In addition, phase-coherent Doppler technique is integrated into this system to determine the location and size of pulsatile structures. This combination of pulse-echo and Doppler techniques also appears promising in detection of myocardial infarctions [42]. The frequency contents of echoes from infarcted and normal myocardium are being studied to determine if the former has a characteristic 'signature' which could be correlated with the state of pathology.

In yet another study, focused ultrasound is being used to create regions of focal necrosis within the myocardium to serve as models of myocardial infarction [42] more reliably and reproducibly than is possible by techniques based on coronary occlusion.

Lung, composed of tiny air sacs presents a different problem. Tissue-Air interface presents a large mismatch of acoustic impedances. In addition, lung has very high acoustic absorption ($\alpha/f =$

$41 dBcm^{-1} MHz^{-1}$). Thus, pulse-echo techniques are not very useful except in detection of pleural thickening; but resonant absorption and sonic scattering techniques under development [43] may prove very useful in early detection of lung disease without the hazards of radiography.

In contrast to such sophisticated systems, it is feasible to use the simple pulse-echo system to detect non-invasively the formation of ice intracellularly or extracellularly in tissues or organs during their freezing for preservation [44] or to use the focused beam to place occluding lesions in the epididyms for non-surgical sterilization of the male [45]. Thus indeed, ultrasound promises many an exciting application in biomedicine. The question of toxicity of ultrasound in its diagnostic applications is still <u>sub judice</u> and will not be completely answered until the mechanism(s) of action of ultrasound--at the frequency, power levels and the modes actually used--are known with less ambiguity [46]. But meanwhile pragmatically it should be considered safe as in histopathological [47, 48:65] and behavioural studies [49:296] no untoward effects have been observed.

ACKNOWLEDGMENTS

This work is supported by U.S.P.H.S. Grant No. NS 08571 and No. 5-SO5-RRO7047-05. Thanks are due to the Editors of the Journal of Physiology, Experimental Neurology, Environmental Biology [50:207] and Academic Press for permission to use previously published data.

REFERENCES

[1]   Wells, P. N. T.   1969.   Physical Principles of Ultrasonic Diagnosis.   London and New York: Academic Press.

[2]   Brown, B. and D. Gordon, eds.   1967.   Ultrasonic Techniques in Biology and Medicine.   Springfield: Charles C. Thomas Publisher.

[3]   Gordon, D. 1964.   Ultrasound: As a Diagnostic and Surgical Tool.   Edinburgh: E. and S. Livingstone Ltd.

[4]   Grossman, C. C., J. H. Holmes, C. Joyner, and E. W. Purnell. 1966.   Diagnostic Ultrasound.   New York: Plenum Press.

[5]   El'piner, I. E.   1964.   Ultrasound - Physical, Chemical and Biological Effects.   New York: Consultants Bureau.

[6]   Gitter, K. A., A. H. Keeney, L. K. Sarin, and D. Meyer, eds. 1969.   Ophthalmic Ultrasound.   St. Louis: C. V. Mosby Co.

[7]   Goldberg, R. E. and L. K. Sarin. 1967.   Ultrasonics in Ophthalmology.   Philadelphia and London: W. B. Saunders Co.

[8]   Kelly, E. ed. 1965.   Ultrasonic Energy.   Urbana: University of Illinois Press.

[9]   Lele, P. P. 1967.   Production of Deep Focal Lesions by Focused Ultrasound-Current Status. Ultrasonics. 5:105.

[10]   Astrom, K. E., E. Bell, H. T. Ballantine, Jr., and E. Heidensleben. 1961.   J. of Neuropathology and Experimental Neurology. 20:484.

[11]   Basauri, L. and P. P. Lele. 1962.   J. of Physiology 160:513.

[12]   Young, G. F. and P. P. Lele. 1964.   Experimental Neurology. 9:502.

[13]   Lele, P. P. Time Course of Development of Ultrasonic Lesion. In preparation.

[14]   Someda, K. and P. P. Lele. An Electron Microscopic Study of Ultrasonic Lesions in the Brain.   In preparation.

[15]   Manlapaz, J. S., K. E. Astrom, H. T. Ballantine, Jr., and P. P. Lele. 1964. Experimental Neurology.10:345.

[16]   Lele, P. P.   To be published.

[17]   Lele, P. P. and J. Hayes. 1964. Conference on Biologic
Effects of Laser Radiation, Boston Laser Conference.   The Institute
of Electrical and Electronics Engineers and Northeastern University.

[18]   Fry, W. J. and F. Dunn.   1962. Physical Techniques in Biological Research.   Vol. 4.   New York: Academic Press, p. 261.

[19]   Lele, P. P. 1962. J. of Physiology. 160:494.

[20]   Lele, P. P. and H. T. Ballantine, Jr.   To be published.

[21]   Ballantine, H. T. Jr., E. Bell, and J. Manlapaz. 1960. J.
Neurosurgery. 17:858.

[22]   Stevens, S. S. ed. 1951.   Handbook of Experimental Psychology.
New York and London: John Wiley and Sons, Inc.

[23]   Kahn, H. A. 1970.   A Theoretical Analysis of Ultrasonic Focusing Systems. Master's Thesis, Massachusetts Institute of Technology.

[24]   Lele, P. P. A Highly Stable, Sensitive Thermocouple Probe
for the Determination of Energy Distribution in Ultrasonic Fields.
In press.

[25]   Hsu, W. L. 1970.   An Improved System for the Coupling of an
Ultrasonic Transducer to the Brain of a Cat or Monkey. Master's
Thesis, Massachusetts Institute of Technology.

[26]   Bellows, A. H. 1962.   Design and Construction of an Optical
Adjunct to Ultrasound Neurosurgical Equipment. Bachelor's Thesis,
Massachusetts Institute of Technology.

[27]   Heuter, T. F. 1956.   Report on Progress on Contract B-816(C)
to the National Institutes of Health, U. S. Public Health Service.

[28]   Lele, P. P. and D. S. Alles. A Radiation Pressure Balance
for the Measurement of Ultrasonic Power. In press.

[29]   Lele, P. P. 1962. J. Acoustical Society of America. 34:412.

[30]   Fry, W. J. and R. Meyers. 1962. Confinia Neurologica. 22:315.

[31]   Hsu, W. L., J. D. McGill, G. A. Wanek, and G. D. Wight.
1966. Project Report No. 2.671. Department of Mechanical Engineering, Massachusetts Institute of Technology.

[32]   Robinson, T. C. and P. P. Lele. An Analysis of Lesion
Development in the Brain and in Plastics by High Intensity Focused
Ultrasound at Low Megahertz Frequencies. J. Acoust. Soc. Amer.
51:1333-1351.

[33]   Fry, W. J. and F. J. Fry. 1963. Anatomical Record. 147:171.

[34]   Lele, P. P. 1963. Experimental Neurology. 8:47.

[35]   Fry, W. J. 1958. Advances in Biological and Medical Physics.
Vol. 4. New York: Academic Press, p. 281.

[36]   Lele, P. P., S. Shibata, and J. Running. 1970. The Physiol-
ogist. 13, No. 3.

[37]   Lele, P. P. 1966. Medical and Biological Engineering. 4:451.

[38]   Matison, G. 1971. Sonar Detection of Ultrasonic Lesions
During and After Production. Doctoral Thesis, Massachusetts Institute of Technology.

[39]   Pandya, D., E. E. Karol, and P. P. Lele. Brain Research.
In press.

[40]   Lele, P. P. and D. Pandya. Unpublished data.

[41]   Lele, P. P., A. D. Pierce, E. A. Woodin, and H. E. Egerton.
Unpublished data.

[42]   Lele, P. P. and J. C. Caulfield. Unpublished data.

[43]   Murphy, R. L. H., E. A. Woodin, and P. P. Lele. Unpublished data.

[44]   Lele, P. P. and E. Cravalho. Unpublished data.

[45]   Lele, P. P. Unpublished data.

[46]   Lele, P. P. and A. D. Pierce. Unpublished data.

[47]   Lele, P. P. Unpublished data.

[48]   Pond, J. B. and R. Warwick. 1970. Megahertz Irradiation of
Pregnant Mice. IEEE Ultrasonic Symposium Abstracts, IEEE Trans.
on Sonics and Ultrasonics. SU-17:65.

[49]   Smyth, M. G., Jr. Animal Toxicity Studies with Ultrasound at
Diagnostic Power Levels. In Diagnostic Ultrasound, C. C. Grossman,

C. Joyner, J. H. Holmes, and E. W. Purnell, eds. New York:
Plenum Press, p. 296.

[50] Lele, P. P. 1966. Environmental Biology. P. L. Altman and
D. S. Dittmar, eds. Federation of American Societies for Experimental Biology, Bethesda, Maryland, p. 297.

CHAPTER 13

EFFECT OF INFRARED LASER RADIATION ON BIOLOGICAL SYSTEMS

George W. Pratt

SYNOPSIS

Within the past year we have found that radiation emitted by the CO_2 laser interacts very strongly with biological systems. Heat resistant spores (Bacillus Subtillus) were rendered completely inactive in 10^{-2} seconds in a 50 watt, unfocused beam, as compared to 3 hours at $170^{\circ}C$ using conventional steam sterilization methods. Similar results have been obtained with steptococcus facium. The characteristic stretching vibration frequencies of many basic organic bonds lie in the 5-12 micron regions of the infrared. Coupling with the laser is very likely by way of these excitations. It may be possible by tuning the laser, controlling pulse powers and lengths, to selectively deliver energy to specific parts of a macromolecule. Preliminary experiments with polyglycine support this concept. We feel that the use of lasers in the infrared will be very useful in the development of new sterilization techniques and possibly of considerable value to the structural chemist in the study of proteins and nucleic acids. This chapter deals with the development of infrared laser sterilization techniques and with a study of mechanisms of molecular alteration.

INTRODUCTION

The effects of electromagnetic radiation on biological systems has been studied for many years. Radiation can be used to enhance biological processes, to inhibit or destroy biological activity, and to induce molecular mutation. The preservation of food stuffs and sterilization of equipment by radiation are familiar applications. This lecture deals with the biological effects and consequent applications of infrared laser radiation.

The electromagnetic spectrum is divided up in Table I showing the type of radiation, typical energies, and the nature of excitation produced in a molecule. Of the various energy ranges only the infrared has been more or less ignored in examining the biological effects of the radiation with the obvious exception of the effects of heating with essentially a black body source. The basic reason for the gap in studying infrared effects is that prior to the laser there has not been

Table 1.

A)	γ - rays	Mev energies	Atomic core electron excitations
B)	X-rays	Kv energies	Atomic core electron excitations
C)	UV-visible	10 to 1 ev	Valence or bonding electron excitations
D)	Infrared	$10^{-1}-10^{-2}$ ev	Vibrational excitations
E)	Microwaves	10^{-4} ev	Rotational and eddy current excitations

a high power, narrow bandwidth source. Now with the CO_2 laser,
for example, one can obtain cw operation in the 100 watt range with
ease and pulses in the megawatt range are readily available.

Before taking up the effect of the infrared laser on biomaterials,
we briefly consider the effects of higher energy radiation. Radiation
inactivation and mutation of nucleic acids and proteins by γ-rays,
x-rays, and UV has been studied for some time [1:272]. γ-rays and
x-rays excite core electrons into high lying unbound levels producing
a positively charged excited species. UV will photoionize the bonding
electrons. When the excited system relaxes to a low energy state,
energies in the several ev range must be taken up by the molecular
system. This can be accomplished by several mechanisms. Covalent
bond rupture and the creation of high energy meta-stable molecular
states are certainly two of them. Phonons are a much less favorable
mechanism since these excitations generally lie at less than 0. 4 ev.
The microscopic effect on the molecule from the de-excitation of the
radiatively excited state is well established in several cases. In
some proteins one finds decarboxylation, deamination, and ring
rupture in amino acid parts, breakage of the peptide bond, and of
special importance, breakage of the disulphide bond in cystine which
alters tertiary protein structure. Nucleic acid-residues are quite
sensitive to UV. Photohydration and photodimerization are found to
occur as well as some transfer of excitation within the excited
molecule.

In a sharp contrast the infrared laser directly excites molecular

vibrations as can be seen from Table I.   In principle these modes in-

volve the entire molecule.   However, the primary feature of a given

mode is usually the vibration of individual bonds [2:118].   Therefore,

if the frequency, power level, and pulse duration of the laser are

correctly controlled, the infrared laser should be very effective in

specific bond breaking or in predetermined structural changes, but

perhaps of considerable practical interest is the situation when the

laser is simply used to produce intense, rapid heating effects leading

to sterilization of the target material.   The infrared radiation will

couple to those bonds or vibrational modes of groups where an oscilla-

ting dipole moment is involved and in which the laser frequency lies

within the line width of the mode.   Typical vibrational widths are 50-

100 cm^{-1}, while the laser line width is roughly 50 MHz or 5 orders of

magnitude less.   Large amplitude oscillations can be induced which

may sever the bond or result in structural changes by way of altering

weaker bonds, principally hydrogen bonds.   These weak bonds are

responsible for the conformation of proteins and nucleic acids in

their biologically active form.   Denaturation of a protein of nucleic

acid due to the disruption of its hydrogen bonding pattern results in

destruction of the biological activity [3].

An example of how a mutation in a nucleotide might possibly be in-

frared induced is described with the help of Figure 9-20 of Watson's

book reproduced here in Figure 1.

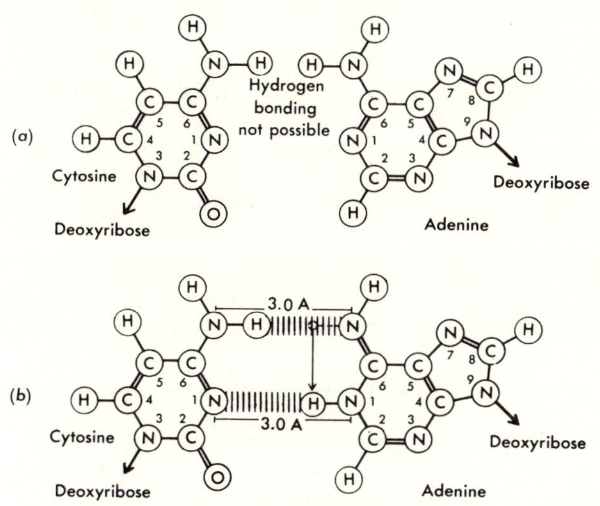

Figure 1.   This demonstrates how the specificity of base pairing in DNA is determined by hydrogen bonding.   The hydrogen atoms are indicated by solid circles, the bonds by ||||||.   (a) shows why the pairing of cytosine with an adenine molecule, having the most stable distribution of its hydrogen atoms, cannot lead to hydrogen bonding.   (b) shows how the shift of a hydrogen atom in an adenine molecule, from the 6-amino group to the N_1 position, permits hydrogen bonding with cytosine.   The normal position of the hydrogen atom is indicated by the small open circle.   (The dimensions shown are only approximate.)

The NH_2 bending mode occurs at 1665 cm^{-1} and is a strong infrared line.   A large oscillation in this mode might lead to a shift from the amino to the imino state of adenine.   In this form, hydrogen bonding to cytosine is possible allowing for a genetic alteration to occur. Löwdin has discussed such mutations in detail [4].

Large amplitude bond oscillations will, of course, make anharmonic terms in the molecular potential important.   This will result in a response at the driving frequency ω_d and at $2\omega_d$ and in the extreme non-linear case at many multiples of ω_d.   In this way very complex excitations of a molecule may be created and it is these indirect couplings that eventually establish a Bose distribution of the excited states as though the molecule has been heated by a black body rather than a very monochromatic source.

Tables II and III [5] give a list of characteristic stretching frequencies near the CO_2 and CO laser emissions respectively. Table IV [6:45] lists some of the modes in nucleic acids in these two regions.

A very attractive possibility is that by controlling the frequency of the laser it may be possible to interact selectively with specific bonds or atomic groups in a molecule and in this way perform selective molecular alterations. The CO laser is particularly adapted to frequency tuning. Aside from the main lines in the 1923 cm^{-1} to 1667 cm^{-1}, there are other recently discovered lines extending all the way to 1111 cm^{-1}. Cw operation, at approximately 10 watts, can be achieved at many of these lines and further fine tuning, using spin flip Raman scattering over roughly 200 cm^{-1} of any particular line, can be carried out [7]. The CO_2 laser has two principal lines, one near 961 cm^{-1} and the other near 1064 cm^{-1}. Another very interesting frequency tuning technique is that of infrared difference frequency generation using a tunable dye laser. Dewey and Hocker [8] have successfully mixed an intense pulse of pump photons at ν_p from a ruby laser with the much less intense output of a dye laser at ν_D in a $LiNbO_3$ crystal to obtain an output at $\nu_p - \nu_D$ whose intensity is proportional to the pump power and inversely proportional to the differ-

Table 2.   Characteristic Stretching Frequencies of Groups in Molecules.

Frequency cm^{-1}	Group	Compound
884-899	(cyclopentane ring)	Cyclopentane and monoderivatives
939-1005	(cyclobutane ring)	Cyclobutane and derivatives
990-1050	(benzene ring)	Benzene and mono to tri-substituted
1020-1075	C-O-C	Aliphatic-compound
1085-1125	C-OH	Aliphatic-compound
1120-1130	C=C=C	Aliphatic-compound

CO_2 laser emits strongly near 961 and 1064 cm^{-1}

Table 3.

Frequency cm^{-1}	Group	Type of Compound
1590-1610	(benzene ring)	Benzene derivatives
1610-1640	N=0	Aliphatic
~1630	C=N	Aromatic
1654-1670	C=0	Aliphatic
1650-1820	C=0	Aliphatic
1695-1715	C=0	Aromatic

CO laser emits strongly near 1510, 1659, and 1790 cm^{-1}

Table 4.

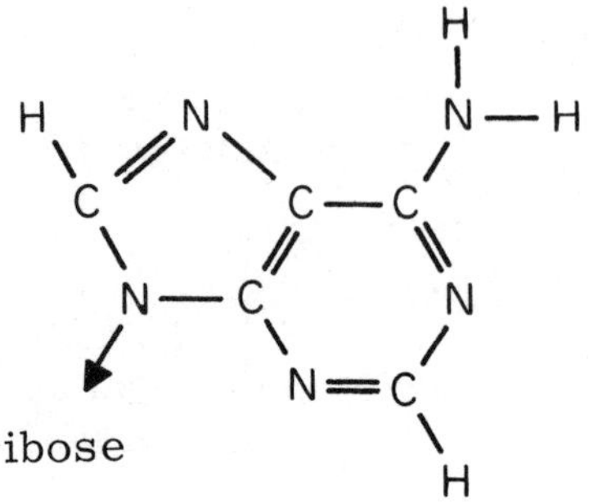

Deoxyribose

Adenine

$$NH_2 \quad 1665$$
$$C=N \quad 1605$$

Uracil

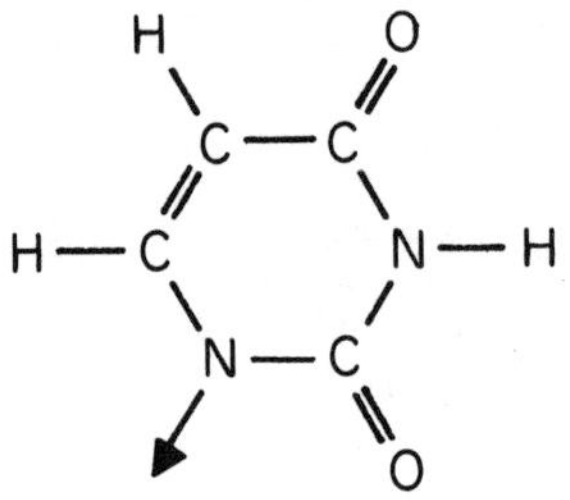

$$C=O \quad 1700$$
$$NH \quad 1680 \text{ cm}^{-1}$$
$$C=O \quad 1650$$

Deoxyribose

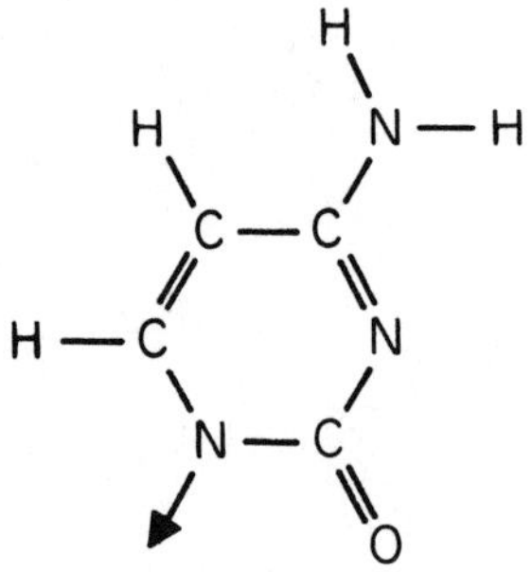

Cytosine

$$1647$$
$$1603 \quad \text{cm}^{-1} \quad \text{In plane ring}$$
$$1585 \qquad\qquad \text{vibrations}$$

---

ence signal wavelength. They obtained infrared signals in the kilo-watt range for pulses of 10 to 20 nanoseconds. This type of approach should permit useful infrared pulses over a very large part of the frequency spectrum of interest for biological systems.

EXPERIMENTAL RESULTS

The CO_2 laser can be used to obtain surface temperatures as high as $5000^{\circ}F$. Metals can be drilled or welded and rock vaporized. It is quite clear then, that the heating effects of the laser can be used to destroy biological activity. As pointed out in the previous section, the great flexibility possible through control of the frequency pulse length, and power levels, offers hope that controlled changes may be effected in the target.

The first experiments were carried out using a very heat resistant dry spore Bacillus Subtillus. These spores were initially spread on Al substrates because metals, such as Al, reflect virtually all of the incident CO_2 laser radiation up to very high incident powers. These spores have a decimal reduction time D (that time required to reduce the number of viable organisms to $1/10^{th}$ of the original value) of approximately 60 seconds in pressurized steam at $130^{\circ}C$. The entire spore population was deactivated in $1/25^{th}$ of a second in an unfocused CO_2 beam covering about 1 cm^2 at 20 watts. This represents an improvement in D by more than a factor of 1500 over the fastest known method of deactivating these spores.

The spores are deactivated so rapidly in the CO_2 beam that it is not necessary to disperse them on a metal substrate. A 20 watt/cm^2 beam pulsed for 1/20th of a second delivers one joule to the target. If all of this energy were absorbed by a paper substrate 10^{-2} cm thick and 1 cm^2 area, the temperature would rise only by 50°C which is well below the rise necessary for ignition. The same energy delivered to a 10 μ thick spore layer would raise the temperature of these spores by roughly 500°C. Assuming approximately equal densities and specific heats the temperature rise of a surface layer δT_L of thickness t_1 on a substrate of thickness t_s is in a very rough approximation given by

$$\delta T_L = \frac{t_s}{t_1} \delta T_s$$

where T_s is the temperature rise of the substrate. Experiments carried out where the spore layer was spread over acetate paper showed that 1/25th of a second pulses at the 20 watt operating level completely destroyed the spore activity, but made no visible change in the paper substrate. This result is of considerable practical importance. It means that surface contamination can be rendered sterile on materials that are quite sensitive to heat and which cannot be sterilized using ordinary techniques, e. g. , steam. The list of such materials is extensive including sutures, plastic and rubber catheters, dressings, etc. In addition many materials are unsuitable for radiosterilization with γ rays. Glass darkness and some rubber

and plastics deteriorate mechanically. Metallic instruments, vessels,

etc. reflect most of the infrared energy until incident power levels in

the $K\omega/cm^2$ range are attained. Organisms on the metal surface

would, however, absorb the laser energy and be destroyed. The CO_2

and CO lasers offer the very attractive feature of sterilization of

metallic surfaces in seconds rather than customary times of several

minutes. Flash sterilization of foodstuffs is a further application

again where surface contamination is involved.

Similar experiments have been carried out using streptococcus

facium. The deactivation obtained in 60 seconds in the CO_2 beam

was the equivalent to that produced by a radiation dose of 2.5 M rad.

This requires 500 minutes from the MIT γ-ray source.

An interesting question is what is the nature of the "heating" pro-

cess in a laser pulse of the order of 10^{-2} seconds duration if all the

energy absorbed were taken up by specific vibrational modes of the

irradiated molecules? Certainly 10^{-2} sec is enormously long com-

pared to vibrational periods and to probable molecular relaxation

times. One can estimate the molecular relaxation times from the

infrared line widths. A line width of 50 cm^{-1} in a 1000 cm^{-1} line

corresponds to a relaxation time of 1.2×10^{-13} sec. Therefore,

energy absorbed in a given molecular vibrational mode is rapidly

dispersed through the excitation spectrum. A 20 watt beam pulsed

for $1/20^{th}$ of a second delivers 5×10^{19} photons to a target area of

1 cm^2. This is approximately 10^5 quanta per absorbing bond per

second. The deviation from the average number of quanta of vibra-

tional excitation is in a specific absorbing mode is roughly

$$\frac{n - n_o}{\tau} = 10^5$$

with $\tau \sim 1.2 \times 10^{-13}$ sec we see that $n - n_o$ is essentially zero. At

$10.6\,\mu$ each photon carries 0.117 ev and 20 of these photons would

be required to break a normal covalent bond. However, a typical

hydrogen bond has a binding energy of 0.1 ev. Consequently at these

low radiation levels from the CO_2 laser it is almost certain that

molecular alterations will be due to disruption of hydrogen bonds

effecting total or partial denaturation.

The estimate above pictures individual molecules with an absorbing

mode treated as a damped oscillator whose line width is described by

a relaxation time τ, which is typically 10^{-12} sec. long. Temperature

is not well defined at times as short as this. In order to invoke

reaction kinetics and heat flow equations enough time must elapse for

the transfer of energy between molecules. This is the thermal

relaxation time τ_h and is related to the thermal conductivity. Typical

values are $10^{-6} - 10^{-7}$ seconds. Hayes and Wolbarsht [9] point out

that laser pulses short compared to τ_h heat up the target before the

energy can be conducted away as heat. The temperature rise in

$$\Delta T = FE_{abs}/C$$

where C is the heat capacity of the target particle, E_{abs} the absorbed energy, and F the fraction of the absorbed energy available for heating. F is given by Hayes and Wolbarsht as

$$F = \frac{\tau_h}{\tau_e}(1 - e^{-\tau_e/\tau_h})$$

where τ_e is the exposure time. In order to convey some feeling for the numbers involved in heat flow we quote a result of Vassiliadis [10] that a melanin granule 1μ in diameter with an initial temperature rise ΔT cools to $1/e$ of ΔT (after the radiation is shut off) in 4×10^{-7} seconds.

The extent of bond breaking due to heating is described by reaction kinetics. The equilibrium constant K_{eq} for a given bond is

$$K_{eq} = \frac{[\text{number of intact bonds }]}{[\text{number of broken bonds}]}$$

and is related to the bond strength ΔG by

$$K_{eq} = e^{-\Delta G/RT}$$

A ΔG typical of a hydrogen bond is -2.726 K cal/mole and it leads to the following ratio as a function of temperature

T^oC	K_{eq}
500	5.6
300	9.8
100	30
25	100

A ΔG of -27.26 K cal/mol is representative of a covalent bond strength and gives a K_{eq} of $10^{7.5}$ at 500^oC. Thus heating a biomaterial to only 300°C breaks 10 percent of the hydrogen bonds but only one covalent bond in 30 million even at 500°C.

It has been stressed above that the infrared laser allows one to deliver large amounts of energy at a specific frequency and over a time domain ranging from cw to pulses of 10^{-12} sec duration. A key question is whether or not this energy can indeed be directed into specific bonds in a protein or nucleic acid. If this is successful it will then be a question of whether controlled molecular changes can be brought about. When the exposure time is long compared to <u>intra</u>-molecular relaxation times, then a given vibrational mode will leak away its excitation energy. The heat energy may be somewhat con-fined in the molecular structure but will more probably feed energy into all of the molecular vibrational and rotational modes. Therefore, the exposure time would seem to be confined to periods of less than 10^{-9} seconds if only one particular bond is to be affected. An ex-posure time between the <u>intra</u> and <u>inter</u> molecular relaxation time will heat the exposed material so rapidly that the heat cannot escape. This will lead to the maximum temperature rise for a given energy input. One concludes then that sterilization will be most effectively accomplished under these conditions.

A step towards the development of pulse techniques has been taken by a senior thesis student, Marshall Schorin. Simple proteins, polyglycine and phenylalanylglycine were pressed in KBr pellets. The infrared transmission spectrum was run on the pellet which was then exposed to pulsed CO_2 radiation and the transmission spectrum run again. Exposure times were long, e. g. , 10^{-1} sec. Therefore, the laser had to act essentially as a heater. Changes in the transmission spectrum of these materials were observed and it differed for different parts of the spectrum. No identification has been made of which particular vibrational modes were most affected. Schorin's [11] results are shown in Figures 2, 3, and 4. The main significance of Schorin's results is that he has established a useful technique that can be applied to higher power, shorter duration pulses.

PROPOSED RESEARCH

This work has been carried out jointly with the Nutrition Department at MIT working with Professor A. Sinsky. We plan to continue examining the sterilization aspects. A variety of spores, phages, and virus samples will be exposed. We plan to look at Bacillus Subtillus in further detail. Yoshikawa and Sueoka [12:806] have found that ultraviolet light induces mutations and have used these results to study the mechanism of replication of DNA in this spore. It would be interesting to compare the ultraviolet mutations with the changes induced in the CO_2 beam.

We further plan to try to make microscopic identification of just how the absorbed infrared energy affects a molecule. Are specific, local changes made, does denaturation occur, are covalent bonds broken. Several means of approaching these questions are available.

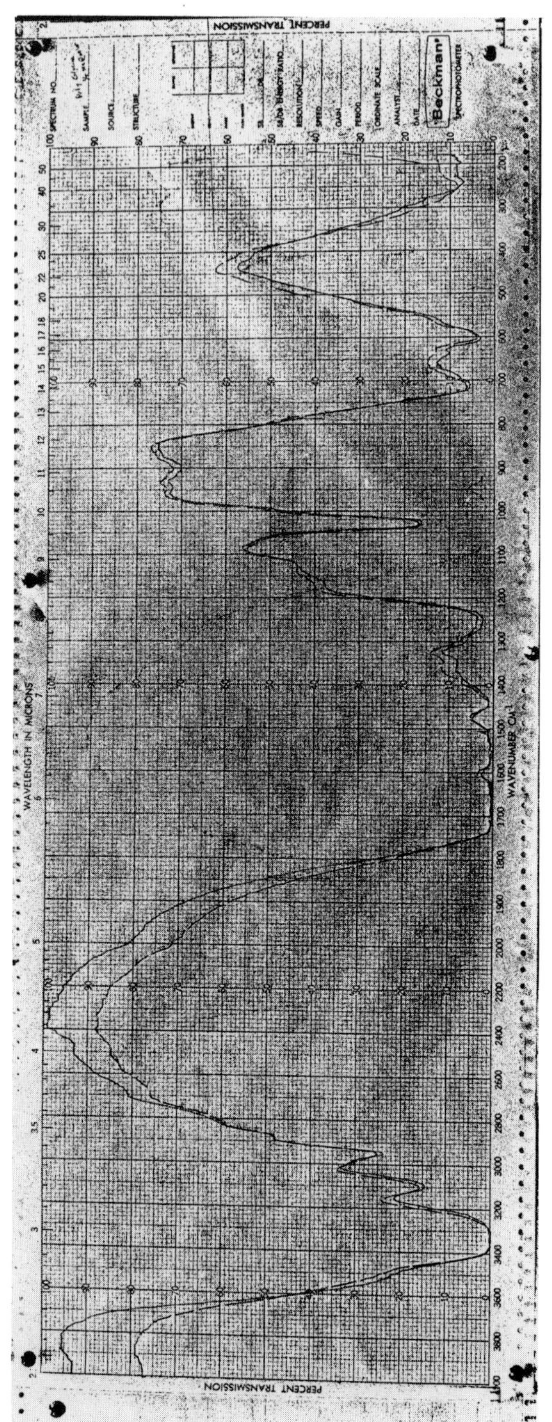

Figure 2. This graph shows the difference between polyglycine unexposed (broken line) and exposed for a short time only (0. 2 sec. at 60 W.) (solid line).

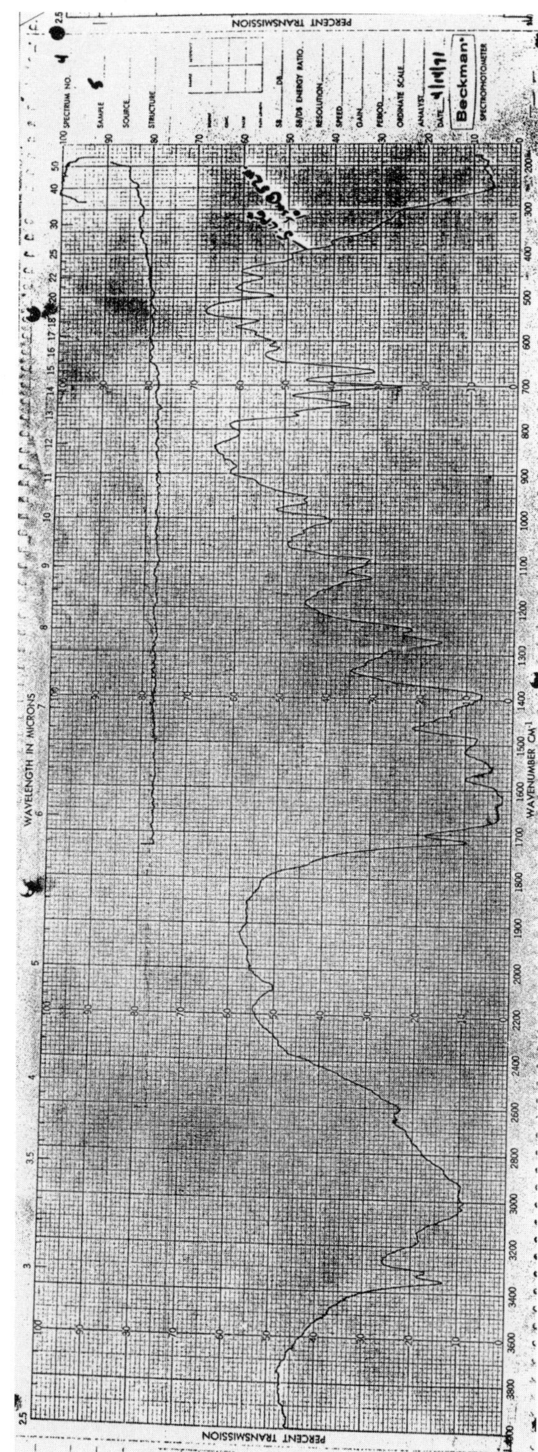

Figure 3. Phenylalanylglycine after small exposure (0.1 sec. at 52 W.). (The top (incomplete) trace was used to check the level of the spectrophotometer; ignore it.)

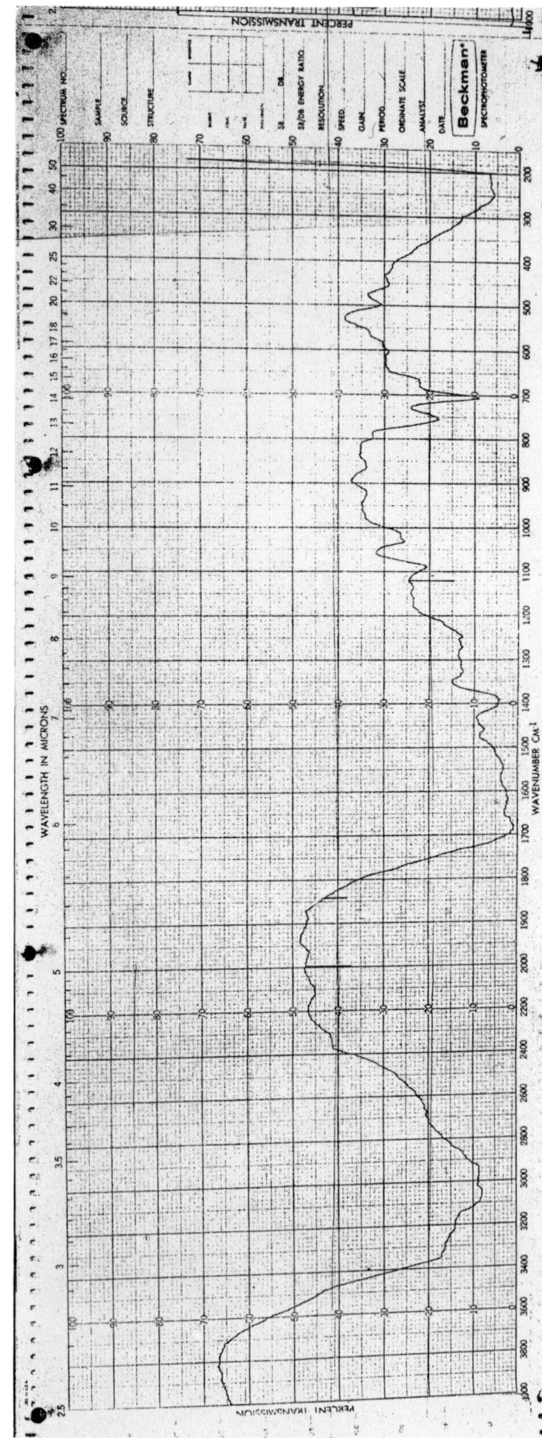

Figure 4.   This graph was the first one to indicate a change in spectrum after irradiation. It is phenylalanylglycine exposed for 0.5 sec. at 60 W.

Schorin' s infrared transmission technique will be pursued. In addition we plan to monitor Raman scattering and optical rotary power during the infrared exposure. The Raman spectra directly gives the vibrational frequencies of the Raman active modes which will be sensitive to a variety of molecular changes. The optical rotary power has been widely used in the investigation of protein and macro-molecular structure. It is sensitive for example to denaturation.

Experiments are now underway which measure the change in optical rotary power induced by a heat spike. Observation of the kinetics of renaturation should be possible as well as a precise determination of the heat input threshold above which the protein cannot recover its original structure. In addition, observations of the fluctuati ons of optical rotary power by photon counting spectroscopy are being carried out. These experiments should allow us to directly observe the dynamic changes in protein or nucleic acid structure and the response to changes in temperature.

REFERENCES

[1] Jagger, J. 1967. Introduction to Research in Ultraviolet Photobiology. Englewood Cliffs, N. J.: Prentice-Hall. Also, Pollard, E. 1959. Rev. Mod. Phys. 31:272.

[2] Sutherland, G. B. B. M. 1959. Rev. Mod. Phys. 31:118.

[3] Watson, J. D. 1965. Molecular Biology of the Gene. New York: W. A. Benjamin.

[4] Lowdin, P. O. Advances in Quantum Chemistry.

[5]   Yariv, A. 1967. Quantum Electronics. New York: Wiley.

[6]   Tsuboi, M. 1968. Applied Spectroscopy Reviews. 3:45.

[7]   Patel, C.K.N. and E.D. Shaw. Applied Physics Letters.

[8]   Dewey, C. F. and L. D. Hocker. Applied Physics Letters.

[9]   Hayes, J.R. and M. L. Wolbarsht. 1971. Laser Applications in Medicine and Biology, ed. M. L. Wolbarsht. New York: Plenum Press.

[10]   Vassiliadia, A., Loc. Cit. Chapter VI.

[11]   Schorin, M. 1971. Senior Thesis. Department of Electrical Engineering, M.I.T.

[12]   Yoshikawa, H. and N. Sueoka. 1963. Proc. National Acad. Sci. U.S. 49:559 and 49:806.

CHAPTER 14

BIOMECHANICS OF REPRODUCTIVE BIOLOGY

Thomas J. Lardner

INTRODUCTION

The term "biomechanics of reproductive biology" used here means the
application of analytical and numerical techniques to the modeling of
different phenomenon in reproductive biology with a view to a better
understanding of physiological functions and related experiments. This
definition of biomechanics of reproductive biology is in accord with the
more general definition of biomechanics as "mechanics applied to
biology," see Fung [1]. In Fung's survey article the scope and history
of biomechanics as well as continuum mechanics problems in physiology are reviewed. It is clear from Fung's article as well as from the
more recent biomechanics literature that there is a wide breadth of
activity in biomechanics, encompassing areas, to name only a few, of
cardiology, biomaterial behavior, microcirculation, and lung physiology.

However, a review of the biophysics and biomechanics literature
shows that with few exceptions very little work has been carried out in
the area of biomechanics of reproductive biology. The exceptions are
the hydrodynamics of swimming spermatozoan, the response of human
uterine muscle [2:254] [3:1055][4:154] and recent work on the bio-

physics of nidation [5] . In addition to the investigations of the stress-

strain properties of strips of human uteri [2:254] [3:1055][4:154],

mechanical properties of the uterus and the vagina of rabbits can be

found in Yamada [6]. Some additional comments on reproductive

biology may be found in [7]. When one further considers the area of

reproductive physiology as a part of general physiology one finds that

it is an area in which there has been comparatively little research

until recently. The reasons for this are many, the most important

probably being the difficulty of experimentation.

A convenient classification of events and processes in reproductive

physiology is shown schematically in Fig. 1. The events occuring in

human reproductive physiology are less numerous and occur on a time

scale much longer than those we associate with cardiology and kidney

physiology. For example, the normal cardiac output of a resting

adult with a heart rate of 72 beats/min is 5. 0 liters/min; the average

volume of fluid filtered from the blood plasma into the Bowman cap-

sules in the kidneys is 180 liters/day. On the other hand, an estimate

of the volume flow rate of fluid in the tubules of the human testis is 1. 5

milliliters/day. The rate of human sperm production per day is

approximately 2×10^8 sperm cells/day -- a seemingly high rate of

production considering that only a single sperm cell plays a role in

fertilization -- is less than one percent of the daily production of red

blood cells needed to maintain a red blood cell count of approximately

25×10^{12} cells in the blood. The major event -- ovulation -- in the

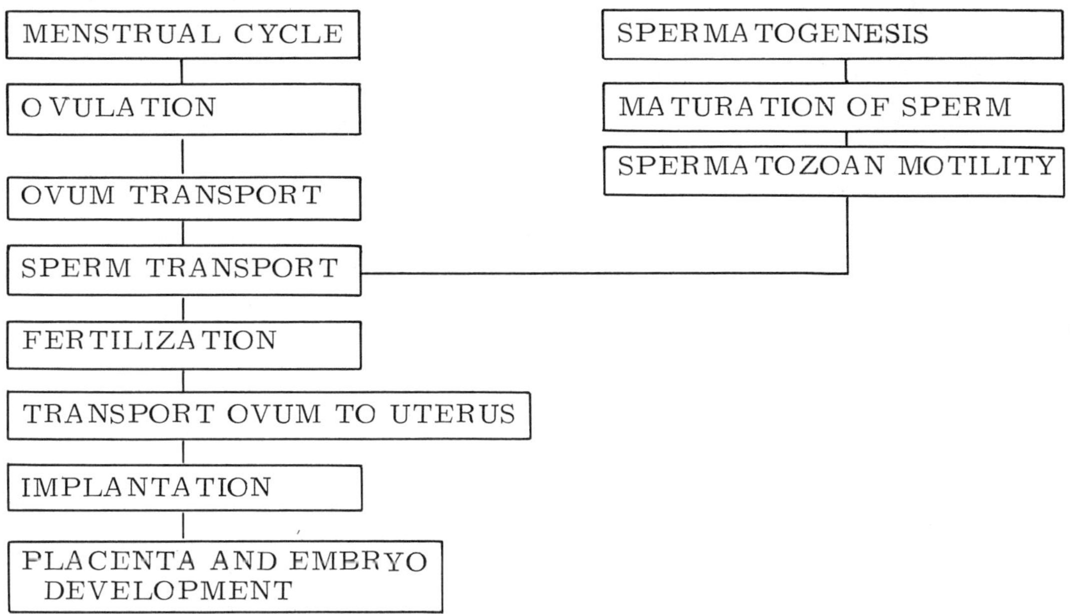

Figure 1.   Schematic diagram of events and processes in reproductive physiology.

female occurs only once a month.

Until recently the events and processes of reproductive physiology were not considered major factors contributing to the health of individuals and/or society as a whole. However, when these infrequent events and what might be considered glacial-like processes are scaled by factors of 10^9 -- the world population -- they take on a greater significance. It is clear also from the concern over the paucity of our knowledge in reproductive biology coupled with a concern over the continued growth of the world population that many different disciplines should be involved in population research [8].

The purpose of this lecture is to survey a number of different problems in the biomechanics of reproductive physiology which are pre-

sently under investigation. The lecture will be descriptive; the details of the solutions for a number of the topics to be discussed can be
found elsewhere. The bibliography has been limited in general to
review papers.

DISCUSSION

Fluid Transport in the Male Tract

An understanding of the mechanisms of fluid transport within the male
reproductive tract must explain first the transport of sperm from the
seminiferous tubules after spermatogenesis to the tail of the epididymis
and second, the transport of sperm from the tail of the epididymis to
the urethra during emission.

Figure 2 shows the different portions of the male tract in their
approximate anatomical arrangement. The vas deferens from each
testis join into the urethra above the prostate gland. Figure 3 shows
in more detail the structure of the portions of the testis. Surprisingly
the detailed structure of the male tract is not well known although more
data continues to become available [9:225][10][11]. The data of interest to an understanding of fluid motion and of diffusion properties
across the epithelium of the different tubules -- cross sectional areas,
lengths, pressures, and flow rates -- are not available. The knowledge of the male tract can be contrasted with that of the kidney for
which details of fluid pressure and volume flow in the kidney nephrons
are known [12].

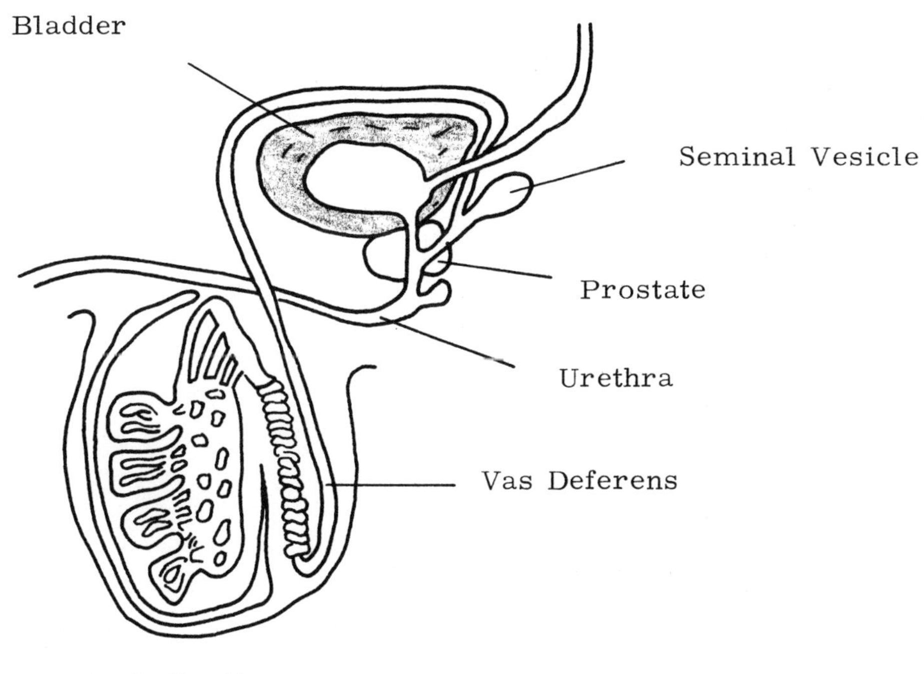

Figure 2.   Male tract.

The process of development of spermatozoa in the seminiferous

tubules -- spermatogenesis -- takes approximately 72 days of develop-

ment from the original germ cells lining the base of the epithelium of

the seminiferous tubules to the release of mature spermatozoa into the

lumen.   On the average approximately 100 million sperm are produced

per day per testis.   In contrast to the cyclic hormonial behavior in the

female, the hormones responsible for spermatogenesis in the male

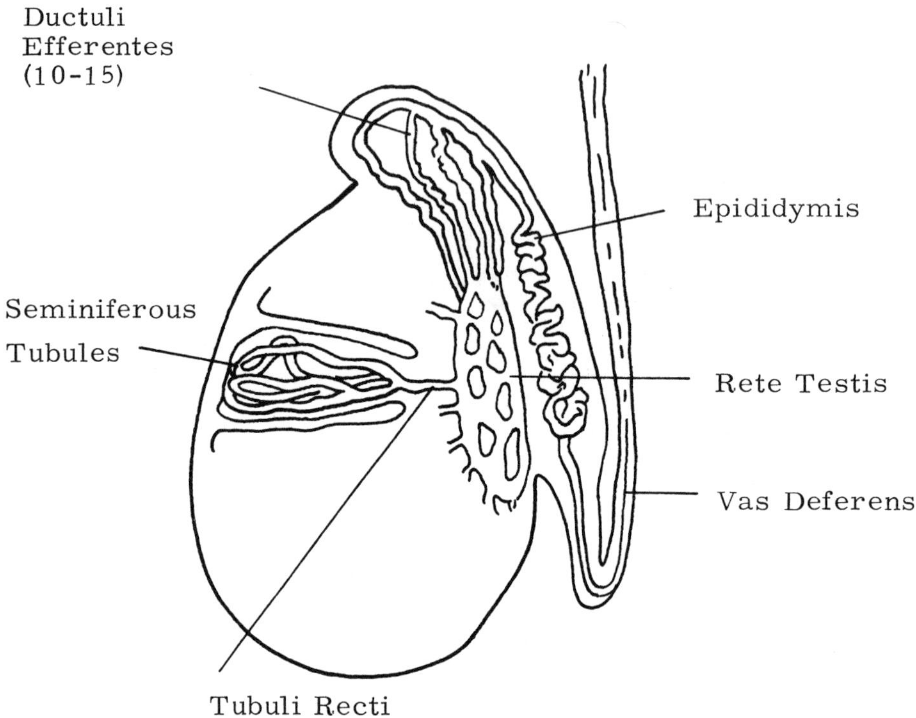

Figure 3.   Structure of the testis.

are released at a relatively constant rate [13][81] and spermatogenesis continues at a constant rate.

The testis is divided into about 250 lobules.   Each testicular lobule contains 1-4 greatly convulated tubules (seminiferous tubules). Each tubule has a length of approximately 70 centimeters and a lumen diameter of approximately 50 microns.   The epithelium of the seminiferous tubules in addition to releasing the spermatozoa into the tubules also actively secretes fluid [9:225][10][11].   This fluid is responsible for carrying the sperm through the various portions of the tract.   The spermatozoa once released into the seminiferous tubules are carried through small collecting tubules into the rete testis. Studies [14:1271] [15:585][16:438]on the structure of the seminiferous tubules show the presence of contractile cells indicating that possibly the tubules can contract to push fluid from the tubules to the rete-testis. The rete-testis is a network of spaces [40]. The fluid from the seminiferous tubules passes into the rete-testis and from there is transported through the approximately 10 to 15 ductuli efferentes, 4-6 cm long, to the head of the epididymis. The epididymis is a highly convoluted tubule 4 to 6 m long.

The ductuli efferentes are of interest because cilia line the epithelium of the tubules; we will discuss the role of cilia in fluid transport below. Cilia are motile, hair-like appendages on certain cells and they may contribute to the transport of fluid from the rete testis into the head of the epididymis [17][40].

At the ends of the ductili efferentes and at the beginning of the epididymis, absorptive cells are present which are responsible for re-sorption of fluid passing through the ductuli efferentes and epididymis [10]. If the ductuli efferentes are ligated near the rete-testis, fluid accumulates causing degradation of the epithelium of the seminiferous tubules. If, however, the ligation is moved to the tail of the epididymis it is found that no fluid accumulates. These experiments indicate that fluid is reabsorbed in the epididymis, (see [40]).

Once the remaining, i. e., non-absorbed, fluid is in the epididymis, the contractile cells and musculature presence in the tissue surround-

ing the epididymis indicate that a peristaltic or musculature contractile pattern moves the fluid through the epididymis into the vas deferens [40].

The vas deferens is a musculature coated tube approximately 40
centimeters in length and having a diameter of the lumen of approximately 0.1 millimeter.

Recent experimental determinations of the flow rates in the rete
testis [9:225][10][11] are (flow rate of rete testis fluid/gm of testis
weight)

in the rat: 22×10^{-3} mℓ/gm-Hr

in the ram: 5 to 10×10^{-3} mℓ/gm-Hr

in the bull: 4×10^{-3} mℓ/gm-Hr

No values are available for humans.   We might on the basis of the
above values select for the flow rate in the human rete testis

5×10^{-3} mℓ/gm-Hr

and use for testis weight approximately 12 gm; thus an estimate of the
flow rate in a human rete testis is

60×10^{-3} mℓ/Hr.

Hence through <u>each</u> of the approximately 10 ductus efferentes we can
assume a flow rate of

6×10^{-3} mℓ/Hr.

It has been estimated [10] that approximately 99 percent of the fluid
leaving the ductuli efferentes is resorbed at the head of the epididymis.

The remaining fluid and spermatozoa then move into the tail of the epididymis by contractile patterns in the epididymis and thence into the vas deferens [40]. The fluid transport in the kidney nephrons should again be mentioned because of the similar quantity of fluid resorption along the nephrons. A very simple diagram of the state of fluid in the male tract is shown in Figure 4.

As shown in Figure 4 the fluid and spermatozoa secreted into the seminiferous tubules pass into the rete testis. The flow rate in the rete testis for the human was estimated above. If we now take this flow rate and neglect for the moment the presence of absorptive cells in the ductuli efferentes we can estimate the pressure drop between the rete testis and epididymis. This estimate of pressure drop obviously depends on a knowledge of the lumen diameter.

A calculation gives approximately 0.20 millimeter of mercury across the tubules. This calculation of the pressure drop between the rete testis and the head of the epididymis indicates that for these small flow rates many different mechanisms could be responsible for fluid transport.

Certainly contractile patterns in the seminiferous tubules will tend to move the fluid into the rete testis and any slight pressure change will then move the fluid into the epididymis.

The time for human sperm transport through the male tract from their release in the seminiferous tubules to their appearance in the ejaculate has recently been determined [18:390]. The shortest transit

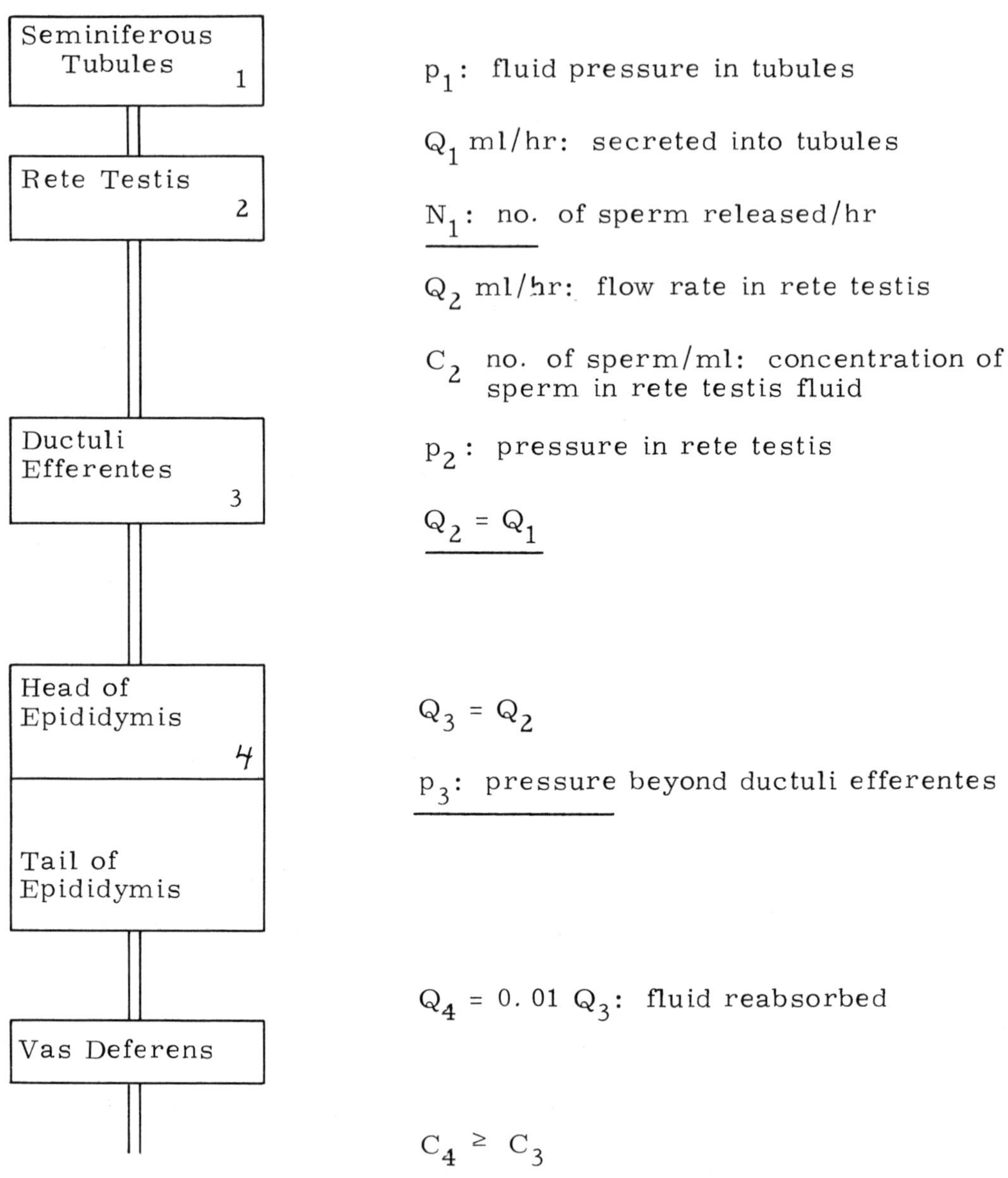

Figure 4.   Fluid flow in the male tract.

time was 1 day, the median transit time was 12 days, and the maximal transit time was 21 days.

The lack of specific data and numbers in Figure 4 beyond those mentioned above emphasizes that the relative contribution of each possible mechanism for the transport of fluid and spermatozoa-absorption, contractile cells, and cilia — needs to be investigated and understood.

One aspect of fluid transport in the male tract that is of particular interest is the role of cilia lining the ductus efferentes between the rete testis and the epididymis. In addition to their presence in the ductus efferentes of the male reproductive tract, they are also present in the oviducts of the female reproductive tract where they play a role in sperm and ovum transport [35:129][36:31][37]. Furthermore cilia are present in almost all groups of the animal kingdom and because of their motility are important not only in reproduction but in locomotion, alimentation, circulation, and respiration [19, 20, 21, 22, 23, 24, 25, 26].

Studies on the hydrodynamics of protozoa which use cilia for locomotion have been made; see for example, the papers by Jahn, [27, 28] and the references therein. A recent paper by Blake [29:199] discusses a spherical envelope model for the swimming of the protozoan _opalina_. A forthcoming paper by Blake [30] considers the swimming motion of either a two-dimensional or cylindrical ciliated body. There has been recent work on the problem of mucus transport in the trachea by the use of a mechanical simulation of the cilia in the trachea, [31:715,

32:309, 33:127]. An analytical model of mucus transport in the trachea
has been presented by Barton and Raynor, [34:419], without consideration of the coordinated wave motion of cilia.

However, while these studies in protozoology and mucus transport
in the respiratory tract have provided some information on the means
of locomotion in protozoa and particle transport in the respiratory
tract, no work to our knowledge has attempted to relate the properties
of the cilia and the coordinated wave motion of ciliated epithelium
(metachronal wave) to the net fluid transport in tubules. In view of
this and for a need of a better understanding of the transport of fluid
through the ductus efferentes of the human male reproductive tract [17]
and the effect of cilia on sperm and ovum transport in the Fallopian
tubes [35:129, 36:31], an analysis of cilia transport in tubules was
investigated [37]. In particular, application of the results for flow
rates in tubules to the experimental flow rates observed in the ductus
efferentes (see above) of the reproductive tract was made.

We will not discuss the analysis here except to note that the formulation follows analysis similar to peristaltic pumping, [38], and gave
results for the flow rates in the ductus efferentes which were of order
of magnitude

$$Q \approx 0.12 \times 10^{-3} \, m\ell/Hr$$

Unfortunately this value is not in agreement with the value estimated
above ($6 \times 10^{-3} \, m\ell/Hr$). However, in view of the approximations which
were made in obtaining numerical values for flow rates and in the

parameters in the model, close agreement cannot be expected. In an order of magnitude sense, however, it appears possible that ciliary action is responsible for fluid transport in the ductili efferentes. Obviously more work in the physiology of this portion of the male tract is needed before stronger conclusions can be made about the role of cilia in fluid transport in the male tract. In addition, further general analysis and experiments are needed on the effect of cilia in fluid and particle transport.

The transport of sperm from the end of the epididymis through the vas deferens during emission is not understood although it is assumed that peristaltic pumping is responsible [39:86]. No information appears to be available on the mechanical properties of the wall of the vas deferens.

Spermatozoan Motility

One of the fascinating aspects about spermatozoa is their inherent motility. It is probably fair to say that since the time of Van Leeuwenhoek's reports in 1677, swimming spermatozoa have continued to fascinate biologists. Leeuwenhoek believed that spermatozoa played a vital role in fertilizing the egg. However, in 1780, Spallanzani concluded that spermatozoa had no such role. By 1890 enough was known about spermatozoa to conclude that its principal function is to initiate the development of the egg by supplying the paternal genetic material. A review of the early history of spermatozoan motility may be found in Bishop and Walton [41] and Bishop [42:1].

Spermatozoa must be equipped with an apparatus for locomotion in order to reach the site of the ovum in the female reproductive tract. It was suggested in 1908 that the longitudinal fibrils within the sperm tail were responsible for spermatozoan locomotion, and later in 1911 that the chemical energy for motility must be distributed locally throughout the length of the flagellum. This is still the currently accepted view [43:775, 44:696, 45:796, 46:88, 47:365].

Only a few experiments <u>quantitatively</u> describing the characteristic motions of spermatozoa were published prior to 1960. Foremost among them have been the works of Gray [43:775, 48:96] and Gray and Hancock [49:802]. A symposium on spermatozoan motility was held in 1960 [50] and since then a number of review articles on sperm movement have been published [42:1, 44:696, 51:27]. Studies on spermatozoa structure and motion have increased significantly due to refinements in techniques and marked advances in instrumentation (electronmicroscopy and high speed cinematography). Even with these innovations in experimental techniques much of the literature of spermatozoan motility is dominated by observations of sea-urchin and other invertebrate sperm. The propulsive mechanism of the spermatozoa of invertebrate species have been investigated by Gray [43:775], Brokow [52:155, 53:161, 54:445], Holwill and Sleigh [55:267], and Machin [45:796, 46:88]. With the exception of the works of Gray [48:96], and Rikmenspoel [47:365] on bull sperm and Swan [56] on ram sperm no quantitative analysis has been reported on mammalian spermatozoa. Swan [56] in a dissertation at the University of Sidney

has presented an extensive comparative study of the propulsion move-

ment and structure of the spermatozoa of the ram and oyster. She

also presents a complete discussion of the history of spermatozoan

motion as well as a general survey of mammalian sperm structure.

In contrast to the literature on quantative descriptions of mammalian

sperm motility, there is much literature on qualitative descriptions,

see for example, [57:571 to 60:471] .

A better understanding of mammalian sperm in general should pro-

vide better understanding of human sperm. An attempt to answer the

question of "Why are we interested in sperm movement?" was given

by Rothschild [63] ; it is of sufficient interest to quote part of it here:

A suspension of spermatozoa is one of the nicest assemblages of cells
with which one could work. . . . they contain interesting compounds and
enzymes to help them in doing their duty toward the egg; parts of them
are like, or are, muscle fibers so that sperm tails may help us to
learn how muscles contract; parts of them are antigenic, which may
help us to find out whether their tails contain actomyosin, something-
myosin or a new 'muscle' protein; they are easy but sometimes
frustrating game for the electron microscopist; they move in an
obvious way, unlike most other cells, and therefore tell us, visually,
if they are alive (always assuming that if they don't move, they are
dead); they are concerned with some of the most important problems
which affect or afflict the world--over-population and sterility. . . .

In view of the importance of an understanding of spermatozoa

motility we are investigating the relation between spermatozoan

motility and structure as well as the relation between sperm maturity

and motility in different portions of the male tract, see Fray [64].

The objective of the work thus far has been two fold. First to obtain

quantitative experimental data on swimming spermatozoa of the rat and

the rabbit (we are now in the process of extending our work to the
study of other species). Second, to use the results of a general
hydrodynamic theory to find propulsive velocities, internal forces,
bending moments, and power dissipation in the tail of the swimming
spermatozoa. Brokaw [54:445] first applied this general hydrodynamic approach to the study of sea-urchin sperm.

Because Fray's recent thesis [64] surveys our work to date it is of
value to briefly describe its contents. The thesis is divided into seven
sections. In section I, mammalian sperm structure is discussed. The
ultrastructural components of the mammalian germ cell are numerous;
attention, therefore, was focussed on those details which are believed
to be relevant to motility.

Section II briefly presents the general characteristics of spermatozoan motility: the theories of propagated bending waves, factors
affecting motility, and qualitative descriptions of "normal" movement.

The next section is concerned with the hydrodynamics of swimming
microscopic organisms. This section is divided into two parts: the
first part discusses G. I. Taylor's [65:447] formulation and its limitation as well as Gray and Hancock's [49:802] method. Gray and
Hancock's analysis is then applied to flagella motion. The second part
presents a general hydrodynamical approach first used by Brokaw
[54:445].

Section IV presents the experimental procedure as carried out at
the Harvard Medical School under the supervision of Dr. Don W.

Fawcett.   The analysis of the data obtained by high speed cinemato-graphy is discussed in Section V.

In Section VI the experimentally observed swimming motions of the rat and rabbit spermatozoa are discussed.   An analytical solution of the hydrodynamic equations is compared with numerical results obtained from the computer program, and with Gray and Hancock's [49:802] solution for a filament of infinite length.   The results from the computer calculations are compared with the experimentally observed swimming motions.

Finally, Section VII discusses the relationship between the experimental observations and the theoretical calculation.   Recommendations are made for future work which are based on the experimental observations and the theoretical calculations.   A discussion of some of the results in the thesis follow.

Figure 5 shows the forces acting on a typical element of a sperm flagella (Gray and Hancock [49:802], Brokaw [54:445]).   Figure 6 shows the swimming motion of a mature rabbit sperm taken from the

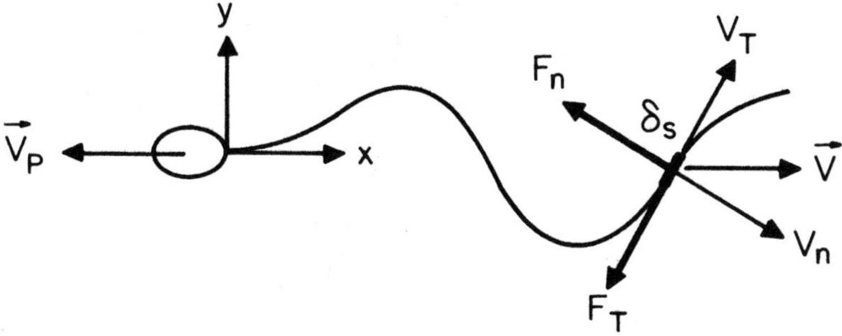

Figure 5.   Forces on an element of a flagellum.

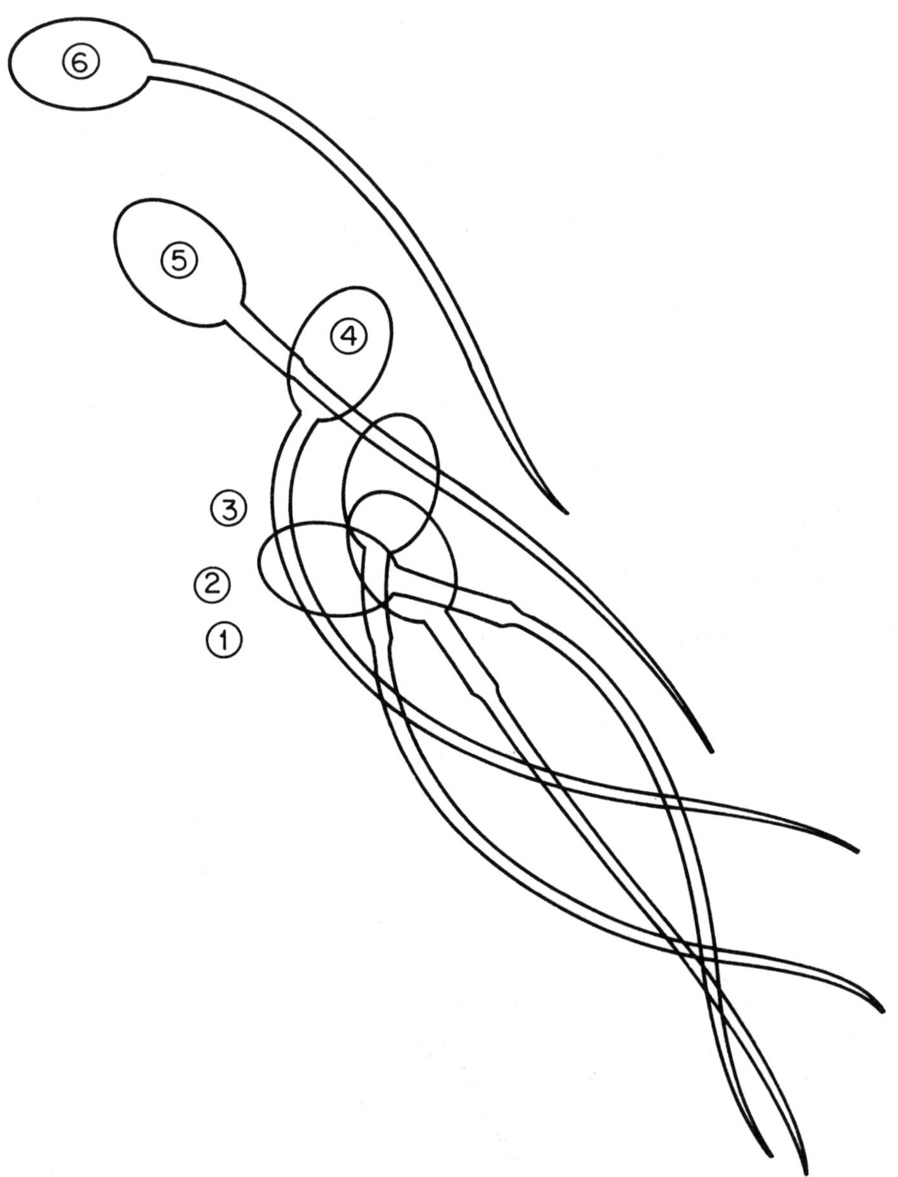

Figure 6.   Tracing of the swimming motion of a mature rabbit sperm.

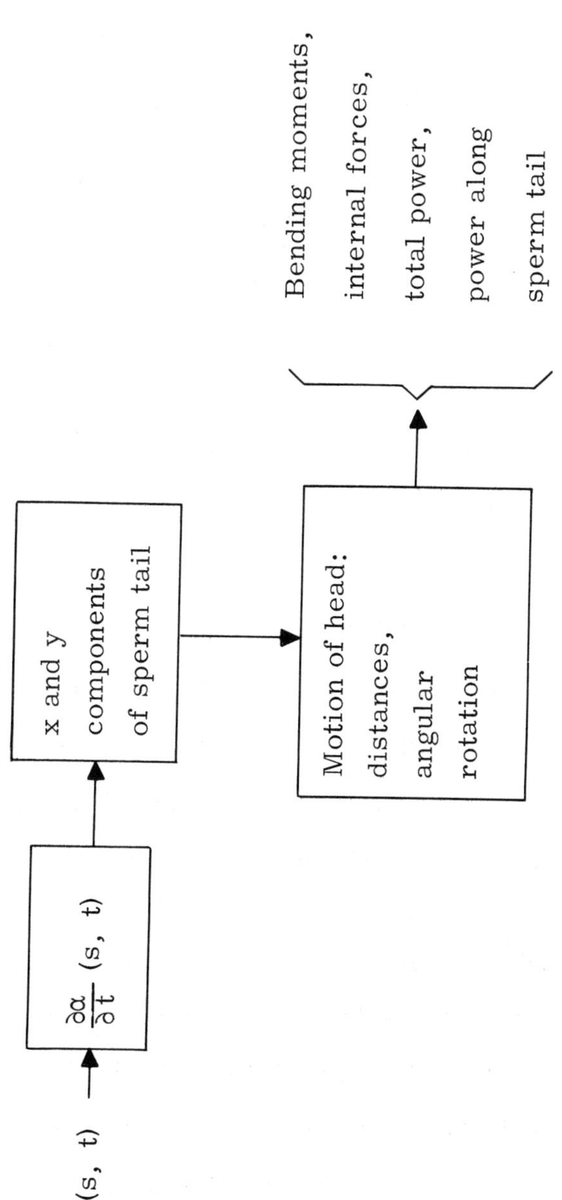

Figure 7. Flow chart of computation for sperm motility study.

tail of the epididymis of the male as traced from the experimental film. The numbers refer to different frame numbers of the film (the time between each numbered sketch is not the same). From tracings of the motion of sperm similar to those shown in Figure 6, it is possible to obtain the tangent angle at different points along the tail relative to a fixed set of coordinate axes. These angles are then used in a computation scheme which is outlined in Figure 7 to obtain bending moments, internal forces, and power expended along the tail. An intermediate step in the calculation allows for the calculation of motion of the sperm which is then compared to the known experimental results. Comparison with experimental values is good. A typical plot for the nondimensional bending moment along the tail is shown in Figure 8.

Fray's work [64] is the first of a series of studies which it is hoped will lead to a better understanding of sperm structure and its relation to motility. In addition we hope to obtain further insight into the maturation process of sperm as they pass through the male reproductive tract from a study of sperm motility in vitro.

Model of the Menstrual Cycle

A possible approach to understanding the menstrual cycle of the human female is to attempt to describe quantitatively the relationships between the participant organs and transient hormone levels in the blood. Until recently most of the experimental work on understanding the menstrual cycle has not been quantitative. This has been a result

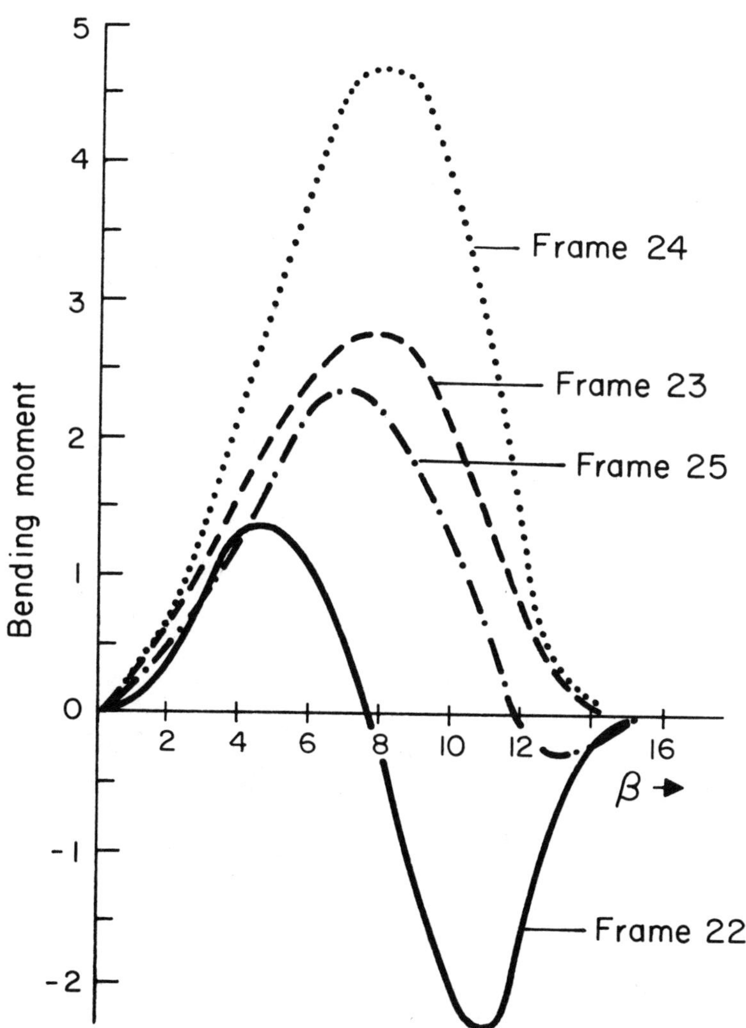

Figure 8.   Calculated nondimensional bending moments along the
flagellum of a mature rat sperm.

of inadequate technology preventing researchers from following the dynamics of the minute bloodstream hormone levels during the cycle. The development of the radioimmunoassay (see [13]) and the protein-binding assay has allowed the precise, relatively easy measurement of hormone levels in small samples of blood (the book cited above provides a description of the theory behind these experimental techniques; see also [81]).

However, even before the development of the radioimmuno- and protein-binding assays, attempts were made to model the menstrual cycle mathematically. Lamport [66:273] first constructed such a model using the "push-pull" theory of estrogen-gonadotropin interaction (blood stream levels of estrogen and gonadotropins are related inversely). This model failed, however, to show the characteristic periodic fluctuation in estrogen levels.

With more data available, and by including the growing follicle explicitly as one of the variables Thompson, et al. [67:278] formulated a model considering FSH, estrogen, and the growing follicle which exhibited sustained oscillations. However, this model is a highly simplified one.

A general description of the interrelationships between the ovary and anterior pituitary in mammals and their cyclic nature in terms of a systems approach was first given by Schwartz [68]. A more detailed model specifically concerned with the estrous cycle of the rat and a discussion of the usefulness of systems theory in endocrine modeling

is presented in [69:1]. A computer simulation of the estrous cycle of
the rat with respect to the blood levels of estrogen and luteinizing hormone and the timing of ovulation is contained in Schwartz and Waltz
[70:1907].

A complete explanation of the human menstrual cycle must include
the interactions of the hypothalamus, the anterior pituitary, and the
ovary. A general overview of the central nervous system-pituitary
ovarian interrelations in women is shown in schematic form in
Figure 9; see also [13]. While the formulation of such a general
model is possible [68] much additional experimental information is
necessary before such a complete model can yield quantitative results.

An attempt to formulate a model of the anterior pituitary-ovarian
interrelationship which includes the interactions between the gonado-
trophins, LH and FSH; the steriods, estrogen, and androgen; and the
growing follicle has been presented by Vande Wiele, Bogumil, et al.
[71]. (This paper also contains an excellent discussion of the physiological mechanisms regulating the menstrual cycle.) The model in
their paper includes only the preovulatory phase of the cycle. It has
been extended to include the complete cycle, [72], and some results
obtained from the complete model are presented in Speroff and Vande
Wiele [73:234], but no details of the formulation of the model are
given.

We have formulated a model for the menstrual cycle which includes
the variables shown in Figure 10, Shack, et al. , [74]. Our primary

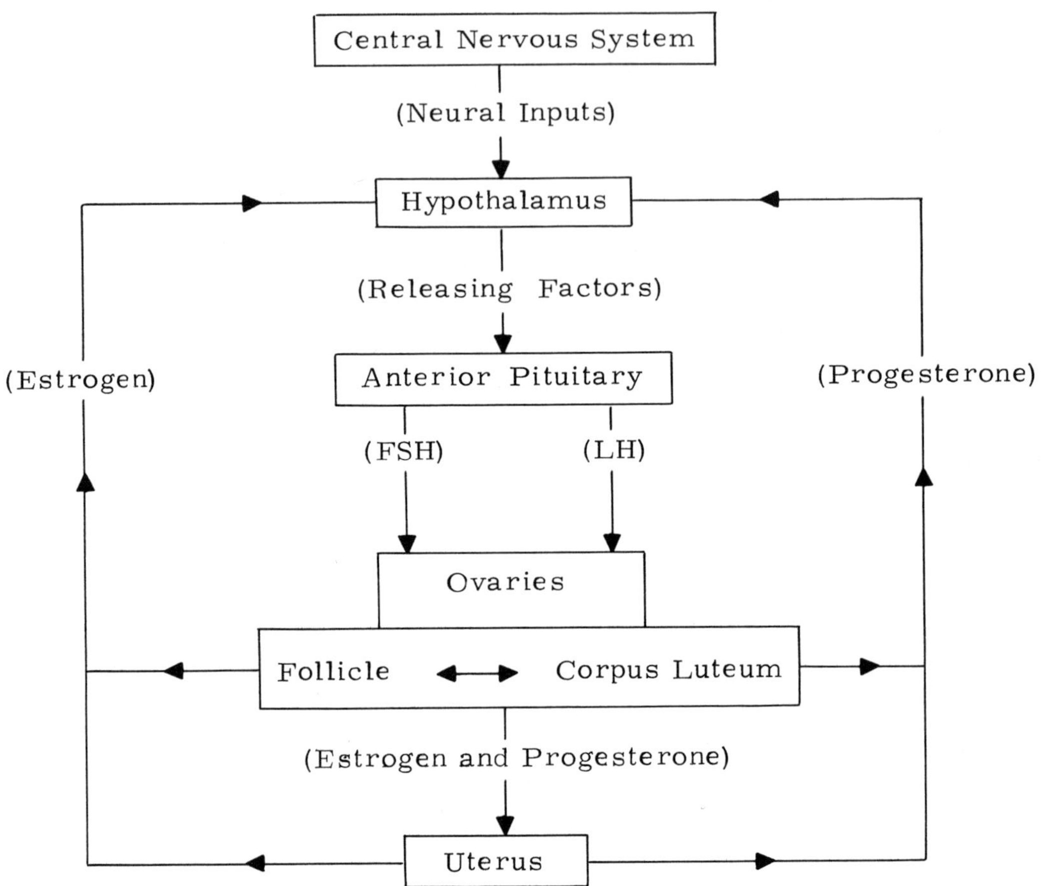

Figure 9.   Mechanisms of human menstrual cycle.

variables are similar to those in [71], but we have made different

assumptions and descriptions of the form of functions describing

steriod feedback, the release of LH and FSH from the pituitary during

the midcycle surge, the transformation of the follicle, and the mech-

anism of ovulation.   Figure 11 shows the flow chart of the model.   It

is hoped that these different descriptions and assumptions will further

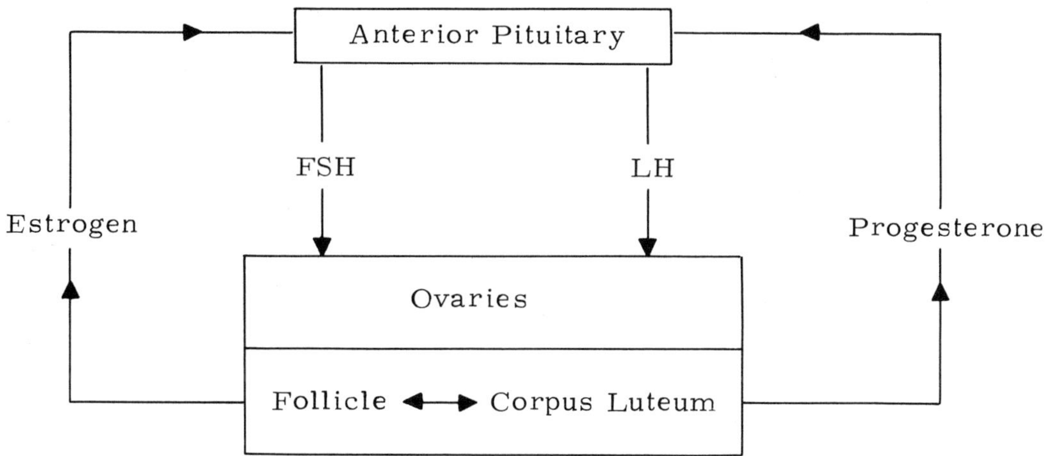

Figure 10.   Mechanisms and variables in the Shack, et al. model.

stimulate attempts to understand actual hormone interactions. The results of the model for the hormone levels are in agreement with measured values [74].

More recent work has attempted to refine and elaborate on the mechanisms considered in the Shack, et al. [74] model providing a better description of different interactions, [75]. In particular, the mechanism and functions describing hormonal clearance, pituitary storage of gonadotropins, the synthesis of progesterone by the corpus luteum, and the mechanisms and conditions for follicle regression were improved.

To further test the hypotheses concerning hormone interactions it was of interest to input the model with exogenous hormone levels similar to those provided by birth control pills, [75].

Birth control pills contain synthetic progestogens and estrogens.

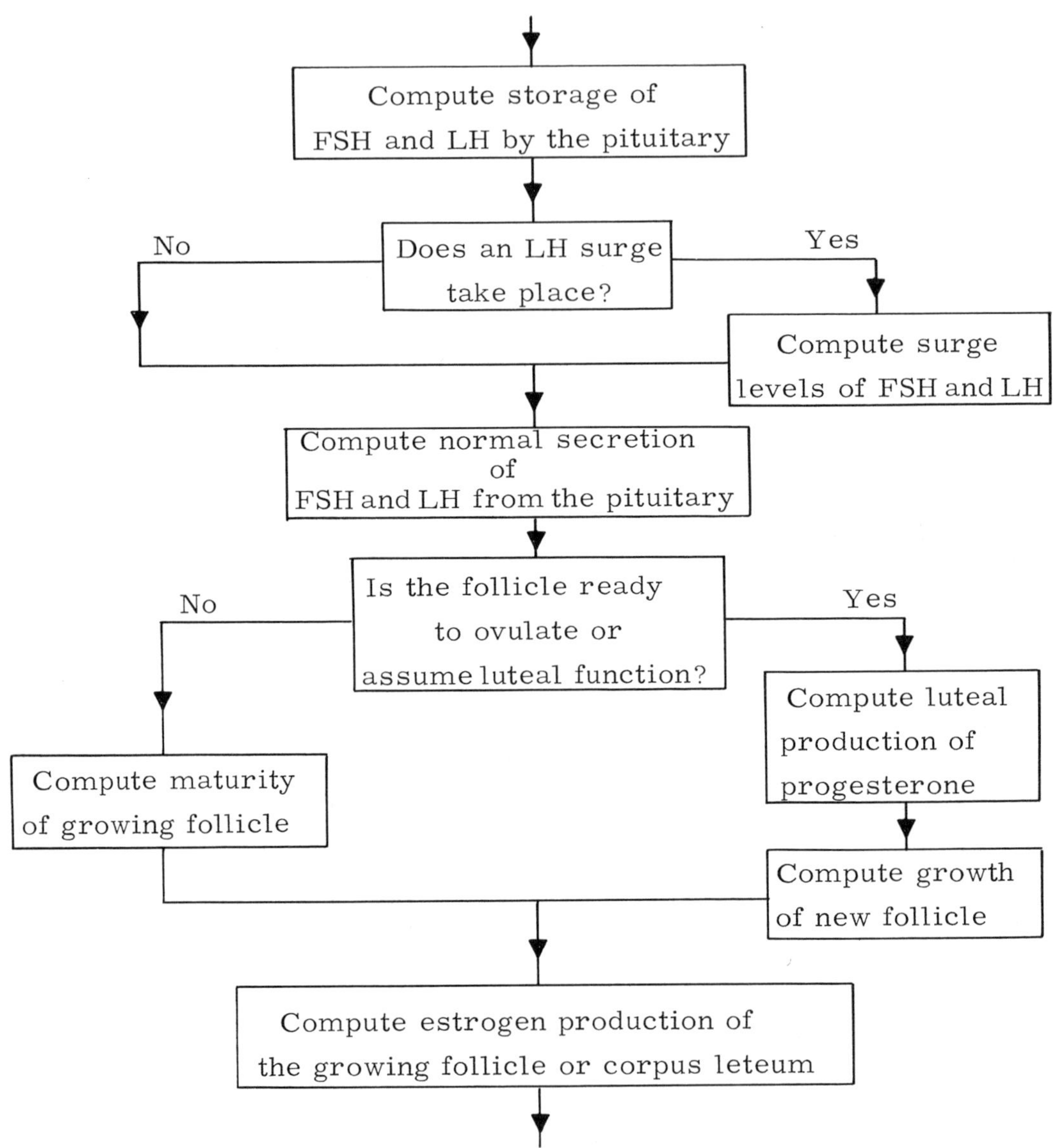

Figure 11.   Flow diagram of the model for menstrual cycle.

These artificial steroids have molecular shapes similar to those of

natural steroids and thus have progestogenic and estrogenic actions similar to that of progesterone and estrogen.

There are two types of birth control pills currently being used, combination and sequential [13]. Combination pills, maintaining high progestogen levels during the period of the cycle when ovulation might occur, are assumed to block the gonadotropin surge and therefore ovulation. They also contain an estrogen that eliminates the side effects of the high progestogen levels. Sequential pills provide high estrogen levels during the period of follicle growth which is thought to suppress the maturation of the follicle until the period of the cycle when the pill provides high progestogen levels (days 22-26). This high progestogen level is again assumed to block the surge and ovulation. Reviews of the speculated mode of action of oral contraceptives are found in [13,76].

A model of the menstrual cycle was run with hormone level inputs similar to those provided by both these types of pills. When preparing inputs to the model, the actual hormonal dosages provided by these pills were altered slightly to compensate for the differences in pro-gestogenic and estrogenic potency between artifical and natural steroids.

Natural steroids are not effective in birth control pills because of their high clearance rates. The model therefore assumes a lower clearance rate for artifical steroids. There is little knowledge as to how quickly artifical steroids are digested and enter the bloodstream and the assumption was made that these steroids enter the circulatory system linearly during a 24 hour period.

The simulation of the combination pill regimen was successful in blocking ovulation. However, in the simulation of the sequential pill regimen the gonadotropin surge was not blocked and ovulation did occur. Further work is obviously required to understand more fully the hormone interactions.

Ovulation

The process of ovulation is an interesting phenomenon because the release of the ovum from the follicle occurs not because of increasing pressure causing the follicle wall to rupture but rather because of decreased strength of the follicle wall; see [77:64-79].

A simple mathematical model for ovulation can be used to show that an increase in the distensibility of the follicle wall permits the radius of the follicle to increase with constant intra-follicular pressure [80]. Further as the radius of the follicle increases the wall thickness decreases, increasing the wall stress until the rupture stress is reached, i. e. , ovulation occurs. The analysis in [80] is a <u>preliminary</u> study of the mechanics of ovulation in which the analysis allows the follicle wall strength to decrease before ovulation. The results show that the hypothesis put forward by Rondell [77:64] can be described quantatively and suggest that further study may be of value.

The analysis is based on the assumption that the follicle initially increases in size by growth of tissue until a time before ovulation; the wall strength then decreases and the follicle continues to grow under

constant pressure until rupture. Rodbard [78:849] in a recent paper
on the mechanics of ovulation presented a model for a description of
ovulation and gives a number of formulae which it is convenient to
use here without further discussion. The model used is a spherical
membrane of constant wall volume with internal pressure; the reader
is referred to Rodbard's paper for further details.

The relation between the internal pressure and radius is given by

$$P = \frac{YW(R-R_o)}{2\pi R_o R^3} \tag{1}$$

where Y = Young's modulus of the follicle wall

    W = Wall volume, assumed constant

    R = Radius of the follicle

    R_o = Reference Radius

In our application of (1) we turn the equation around and ask for the
decrease in the wall modulus to maintain constant pressure:

$$Y = \frac{2\pi R_o^3 px^3}{W(x-1)} \tag{2}$$

where $x = R/R_o$.

As the radius increases, we find from eq. (2) that the modulus of
elasticity initially decreases (distensibility increases). The order of
magnitude of the drop in the modulus from the initial value can be
chosen to be in approximate agreement with the reduction in modulus
of follicle strips incubated in follicular extracts, Rondell [77:64].

Further as the radius increases with increasing distensibility, the
stress in the follicle wall increases.   The variation in wall stress
with radius is given by:

$$\sigma = \frac{pR}{2t} = \frac{2\pi pR^3}{W} \tag{3}$$

for constant wall volume.   Rupture occurs when the wall stress in the
follicle is given by the rupture stress.   It is possible to account for an
observed threefold decrease in modulus over a radius range such that
the wall stress has increased to the rupture value in this range, [80].
This decrease seems to be in agreement with the results of Rondell
[77:64].   Further the analysis has been used to describe the change
in modulus and wall stress with radius and with time in rabbit folli-
cles, Lipner [98], and the results are in agreement with experiments.

Sperm Transport in the Female Tract

Once spermatozoa are deposited in the female tract they begin their
journey to the site of fertilization in the Fallopian tubes; see [82] for
a bibliography on sperm transport and [83 to 89:831].   A study of
sperm transport in the female reproductive tract is concerned with
attempting to answer two basic and related questions:

(1) Why are so many spermatozoa deposited in the female reproductive tract? and

(2) How do the sperm travel through the tract to reach the site of
fertilization in the Fallopian tube.

The number of sperm deposited on the cervix of the uterus is approximately 200-500 x 10^6. Of this number approximately 10-100 are present in the Fallopian tubes. What happened to the others on the journey to the Fallopian tubes?

If we take 50 μ/sec as the translational velocity of the sperm and use 150 mm as the distance to be travelled in the female tract, we find it takes 50 minutes for sperm to travel the distance on the basis of their own motility. However, there are difficulties with this calculation: the velocity of spermatozoa in cervical mucus is closer to 3-50 μ/sec; the spermatozoa are not always travelling in a straight line journey; it is assumed that the spermatozoa start on their journey immediately. These difficulties are mentioned because Rubenstein et al. [86:15] detected spermatozoa at the end of the human Fallopian tubes within 30 minutes after artifical insemination. It follows there-fore that spermatozoa are able to make the transit in less than 30 minutes; a time which does not seem possible by sperm motility alone. Further in-vitro experiments of sperm migration in excised human and bovine genital tracts have been performed [87:118, 88:73]. In these experiments it was found that spermatozoa migrated through the cervical canal but not beyond the internal opening in the uterus (internal os). None were recovered from the uterus or the oviducts in four human specimens with mucus in the cervical canal or in the bovine specimens [87:118]. In the case of cows it has been demonstrated [90:131] that bovine spermatozoa are transported from the cervix to the end of the oviducts in 2 to 4 minutes both after intracervical

artifical insemination and following natural mating. This time is to
be contrasted to approximately 90 minutes on the basis of sperm
motility alone; see further discussions in [88:73].

When one considers the literature on sperm transport in the human
female tract, contradictions abound and it can be safely said that the
exact mechanism responsible for sperm transport is not fully understood. However, it should be clear from the comments above that an
interaction between the contractions and musculature activity in the female tract and the spermatozoa must be responsible for the transport
mechanisms. The exact role of the contractile patterns in the uterus
and in the Fallopian tubes is not clear. However, they must play a
role as our previous remarks indicate; it is how the contractile
patterns contribute to sperm transport that is not understood. Work
underway by Fromm [91] using telemetry systems should give valuable
information on the in-vivo contractile patterns of both the uterus and
Fallopian tubes. We will not discuss the literature on uterine and
oviductal motility in detail here. Instead we will present a number of
simple calculations for sperm transport in the female tract. These
calculations are very approximate and probably can only suggest
possible answers to our original questions.

In view of the large number of spermatozoa in the semen, it can be
assumed that the transport of spermatozoa into the cervix is a random
walk or in the limit a diffusion problem [92:461]:

$$D \nabla^2 N = \frac{\partial N}{\partial t} \tag{4}$$

where N = number of sperm/unit volume and D is an equivalent diffusion coefficient of order of magnitude $25 \times 10^{-4} \, mm^2/sec$, [92:461].
We will consider the configuration shown in the sketch [92:461].

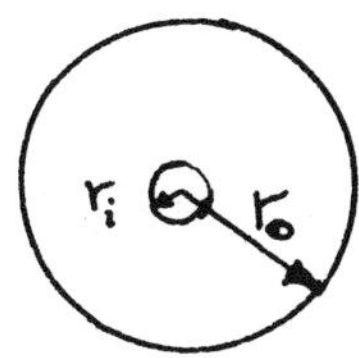

r_i = 1. 25 mm

r_o = 25 mm

The initial condition is $N(r, o) = 1$ and the
boundary conditions are $N(r_i, t) = N(r_o, t) = 0$.
We wish to find the total number of sperm
crossing the inner boundary as a function of
time. However, spermatozoa are only capable
of maintaining their fertilization capacity for
approximately one hour on the cervix; this factor was not considered
in [92:461]. Therefore we wish to know the total number which cross-
ed in one hour. An order of magnitude calculation for the solution of
the above problem gives 1 percent of the initial number moving into the
cervical canal or $1-2 \times 10^6$.

The transport of spermatozoa through the cervical canal is controlled by the cervical mucus; see [93:1] for a complete discussion
and bibliography. The cervical mucus behaves differently during the
menstrual cycle; except for the time near ovulation the cervical mucus
is an amorphorous fluid not amenable to penetration by sperm. However, approxiamtely one or two days preceeding the time of ovulation,
it takes on a different character and becomes laced with long chain
protein molecules. A theory at present [93:1] is that the sperm are
capable of swimming along the channels formed by these long-chained
molecules in the cervical mucus. The presence of these channels on

the hydrodynamics of sperm motility needs to be investigated in addition to the physical characteristics of the fluid.

If we assume that the journey in the cervical mucus causes a drop
by approximately a factor of 10 in the number of sperm which have
entered the cervical mucus, then approximately a hundred thousand to
one million are at the end of the cervical canal.

Once the sperm are in the body of the uterus we need to explain
their subsequent motion. From the in-vitro experiments on excised
uteri mentioned earlier we know that sperm do not swim to the upper
portions of the uterus. It must follow that the contractions and movements of the uterus transport sperm in the direction of the oviducts.
There have been numerous experiments and investigations of the
motility of the human nonpregnant uterus, both in-vivo and in-vitro,
[91, 95:115, 96:441, 97:104]. These experiments provide data on the
rate of contractions, pressure amplitudes, observations on peristaltic
waves and the like. However, these experiments do not explain <u>how</u>
the contractile patterns of the uterus contribute to the transport of
sperm. If we assume that uterine motility moves the sperm to the
upper portion of the uterus, we can ask the question: How many of
these will move into the oviducts?

We might assume as we did above that if 100-200 million sperm
were deposited on the cervix, approximately 100,000 might be found
over the upper 225 mm^2 of the fundus soon after. We now ask the
question how many of these sperm would <u>diffuse</u> into the oviduct open-

ing of a size $\approx 10^{-2}$ mm^2. A diffusion order of magnitude calculation similar to that given above gives about 100 sperm in ten minutes.

It is tempting to answer the question raised by Hartman (quoted in [85], p. 71): "It is however, a mystery still why a 100 x 10^6 sperms should be ejaculated only to die and be absorbed by the female organism," with the observation that unless such numbers are deposited, the number diffusing into the oviducts will be small. We emphasize, however, that these diffusion calculations are only suggestive of the mechanisms responsible for sperm transport; they certainly need further experimental confirmation.

Once the sperm have entered the oviducts the possible mechanisms which might lead to transport up the tubes are many; see [35:129]. We can only mention them here and note that little biomechanics analysis to our knowledge has been applied to their elucidation: cilia transport, motility patterns (peristaltic, abovarian), and Parker's theory [94:381].

The need for data on the oviducts is obvious: lumen sizes, oviductal motility patterns, transit times, oviductal fluid properties, pressure differentials, to name only a few, are the data needed.

CONCLUDING REMARKS

This lecture has attempted to outline a number of biomechanics problems associated with reproductive biology; in a sense it is partially a report of work in progress. It should be clear, however, from the questions raised and from the discussion of the results that there are still a large number of problems yet to be solved. We certainly have not begun to understand many of the events and processes shown in Fig. 1.

ACKNOWLEDGMENT

The work presented here was supported by a grant from the Ford Foundation and the National Science Foundation.

REFERENCES

[1]   Fung, Y.C. 1968. Biomechanics. Applied Mechanics Reviews, 21:1-20.

[2]   Conrad, J. T., W. K. Kuhn, and W. L. Johnson. 1966. Stress Relaxation in Human Uterine Muscle. Am. J. Obs. and Gyn. 95:254-265.

[3]   Conrad, J. T., W. L. Johnson, W. K. Kuhn, and C. A. Hunter. 1966. Passive Stretch Relationships in Human Uterine Muscle. Am. J. Obs. and Gyn. 96:1055-1059.

[4]   Conrad, J. T., and W. K. Kuhn. 1967. The Active Length-Tension Relationship in Human Uterine Muscle. Am. J. Obs. and Gyn. 97:154-160.

[5]   Conrad, J. T. 1970. The Biophysics of Nidation, In Biology of the Blastocyst, ed. J. R. Blandau, Chicago: U. of Chicago Press.

[6]   Yamada, H. 1970. Strength of Biological Materials. Baltimore: Williams and Wilkins.

[7]   Lardner, T. J., W. J. Shack and E. M. Waibel. 1970. A Survey of Reproductive Biology. Report submitted to The Pathfinder Fund, Department of Mechanical Engineering, MIT.

[8]   Prager, D. J. 1970. The Role of Physical Sciences in Population Research. 23rd Annual Conf. on Eng. in Medicine and Biology (ACEMB), Washington, D. C. , p. 81 of proceedings.

[9]   Tuck, R. R. , B. P. Setchell, G. M. H. Waites, and J. A. Young. 1970. The Composition of Fluid Collected by Micropuncture and Catherization from the Seminiferous Tubules and Rete Testis of Rats. Pflugers Arch. 318:225-243.

[10]   Waites, G. M. H. and B. P. Setchell. 1969. Some Physiological Aspects of the Function of the Testis, In The Gonads. ed. K. W. McKerns, New York: Appleton.

[11]   Setchell, B. P. 1970. Testicular Blood Supply, Lymphatic Drainage, and Secretion of Fluid, In The Testis, eds. A. D. Johnson, W. R. Gomes, N. L. Vandemark, New York: Academic Press.

[12]   Hamburger, J. et al. , eds. 1968. Nephrology. Philadelphia: W. B. Saunders.

[13]   Odell, W. D. and D. L. Moyer. 1971. Physiology of Reproduction. St. Louis: C. V. Mosby Company.

[14]   Ross, M. H. and I. R. Long. 1966. Contractile Cells in Human Seminiferous Tubules. Science 153:1271-2.

[15]   Mason, K. E. and S. L. Shaver. 1952. Some Functions of the Caput Epididymis. Ann. N. Y. Acad. Sci. 55:585-593.

[16]   Clermont, Y. 1958. Contractile Elements in the Limiting Membrane of the Seminiferous Tubules of the Rat. Expt. Cell. Res. 15:438.

[17]   Greep, R. , ed. 1966. Histology. New York: McGraw-Hill.

[18]   Rowley, M. , F. Teshima and C. Heller. 1970. Duration of Transit of Spermatozoa through the Human Male Ductular System. Fert. Steril. 21:390.

[19]   Gray, J. 1928. Ciliary Movement. New York: Academic Press.

[20]   Lucas, A. M. 1932. Ciliated Epithelium. Special Cytology. 1: 409-473.

[21]   Faure-Fremiet, E. 1961. Cils Vibratiles et Flagelles. Biol. Rev. 36:464-36.

[22]   Fawcett, D. 1961. Cilia and Flagella. Chap. 4 of The Cell, ed. J. Brachet and A.E. Mirsky, New York: Academic Press.

[23]   Sleigh, M.A. 1962. The Biology of Cilia and Flagella. New York: MacMillan.

[24]   Rivera, J.A. 1962. Cilia, Ciliated Epithelium, and Ciliary Activity. New York: Pergamon Press.

[25]   Satir, P. 1965. Structure and Function in Cilia and Flagella. Protoplasmatologia, IIIE.

[26]   Sleigh, M.A. 1968. Patterns of Ciliary Beating, In Aspects of Cell Motility, Soc. Expl. Biol. Symp. XXII. New York: Academic Press, p. 131-150.

[27]   Jahn, T.L. and E.C. Bovee. 1965. Movement and Locomotion of Microorganism, Ann. Rev. Microbiology, 19:21-58.

[28]   Jahn, T.L. and E.C. Bovee. 1967. Motile Behaviour of Protozoa, In Research in Protozoology, New York: Pergamon Press, p. 40.

[29]   Blake, J.R. 1971. A Spherical Envelope Approach to Ciliary Propulsion. J. Fluid Mech. 46:199-208.

[30]   Blake, J.R. 1970. Private Communication.

[31]   Miller, C.E. 1966. An Investigation of the Movement of Newtonian Liquids Initiated and Sustained by the Oscillation of Mechanical Cilia, Proc. 5th Cong. Appl. Mech., 715-720.

[32]   Miller, C.E. 1967. An Investigation of the Movement of Newtonian Liquids Initiated and Sustained by the Oscillation of Mechanical Cilia, Aspen Emphysema Conf., 309-21.

[33]   Miller, C.E. 1969. Stream Lines, Steak Lines and Particle Pathlines Associate with a Mechanically-induced Flow Holomorphic with the Mammalian Mucociliary System. Biorheology. 6:127-135.

[34]   Barton, C. and S. Raynor. 1967. Analytical Investigation of Cilia Induced Mucous Flow. Bull. Math. Biophysics. 29:419-428.

[35]   Blandau, R.J. 1969. Gamate Transport--Comparative Aspects, In The Mammalian Oviduct. Hafez, E.S. and Blandau, R.J., eds. University of Chicago Press, p. 129-162.

[36]   Sturgis, S. H. 1947. The Effect of Ciliary Current on Sperm Progress in Excised Human Fallopian Tubes. Trans. Amer. Soc. Study Sterility. 3: 31-39.

[37]   Lardner, T. J. and W. J. Shack. Cilia Transport. Submitted for publication, 1971.

[38]   Jaffrin, M. Y. and A. H. Shapiro. 1971. Peristaltic Pumping. Annual Reviews of Fluid Mechanics.

[39]   Mitsuya, H. , J. Asac, et al. 1960. Application of X-ray Cinematography in Urology: I Mechanism of Ejaculation. J. Urology. 83: 86-92.

[40]   Risley, P. L. 1963. Physiology of the Male Accessory Organs. Chap. II of Mechanisms Concerned with Conception, ed. C. G. Hartman. New York: MacMillan.

[41]   Bishop, M. W. H. and A. Walton. 1960. Spermatogenesis and the Structure of Mammalian Spermatozoa, In Marshall's Physiology of Reproduction, ed. , A. S. Parkes, 1:1-129. Boston: Little, Brown and Company.

[42]   Bishop, D. W. 1962. Sperm Motility. Physiol. Rev. 42:1-59.

[43]   Gray, J. 1955. The Movement of Sea-Urchin Spermatozoa. J. Exp. Biol. 32:775-801.

[44]   Holwill, M. E. J. 1966. Physical Aspects of Flagellar Movement. Physiol. Rev. 46:696-785.

[45]   Machin, K. E. 1958. Wave Propagation Along Flagella. J. Exp. Biol. 35:796-806.

[46]   Machin, K. E. 1963. The Control and Synchronization of Flagellar Movement. Proc. Roy. Soc. B185:88-104.

[47]   Rikmenspoel, R. 1965. The Tail Movement of Bull Spermatozoa. Observation and Model Calculations. Biophys. J. 5:365-392.

[48]   Gray, J. 1958. The Movement of Spermatozoa of the Bull. J. Exp. Biol. 35:96-108.

[49]   Gray, J. and G. J. Hancock. 1955. The Propulsion of Sea-Urchin Spermatozoa. J. Exp. Biol. 32:802-814.

[50]   Bishop, D. W., ed. 1962. Spermatozoan Motility. AAAS,

Washington, D. C.

[51]   Nelson, L. 1967. Sperm Motility, In Fertilization. Comparative Morphology Biochemistry and Immunology. Metz, C. B. and Mouroy, A. , eds. New York: Academic Press, p. 27-97.

[52]   Brokaw, C. J. 1965.   Non-sinusoidal Bending Waves of Sperm Flagella. J. Exp. Biol. 43:155-169.

[53]   Brokaw, C. J. 1966.   Bend Propagation Along Flagella. Nature. 209:161-163.

[54]   Brokaw, C. J. 1970.   Bending Moments in Free Swimming Flagella. J. Exp. Biol. 53:445-464.

[55]   Holwill, M. E. J. and M. A. Sleigh. 1967.   Propulsion of Hispid Flagella. J. Exp. Biol. 47:267-276.

[56]   Swan, M. A. 1971.   A Comparative Study of the Propulsion, Movement and Structure of Spermatozoa of a Mammal (Ovis Aries) and a Mullusc (Ostrea Commericialis). Ph. D. Thesis, University of Sydney, Australia.

[57]   Blandau, R. J. and R. E. Rumery. 1964. The Relationship of Swimming Movements of Spermatozoa to Their Fertilizing Capacity. Fert. and Ster. 15:571-579.

[58]   Blokhuis, E. W. 1961.   Optical Investigations on the Movement of Bull Spermatozoa. Proc. 4th Intern. Congress on Animal Reprod. 2:243-248.

[59]   Branham, J. M. 1969.   Movements of Free Swimming Rabbit Spermatozoa. J. Reprod. Fert. 18:97-105.

[60]   Gaddum, P. 1968.   Sperm Maturation in the Male Reproductive Tract: Development of Motility. Anatomical Record. 161:471-482.

[61]   Rikmenspoel, R. , G. van Herpen and P. Eijkhort. 1960. Cinematographic Observations of the Movements of Bull Sperm Cells. Phys. Med. Biol. 5:365-392.

[62]   Zorgniotti, A. W. , R. S. Hotchkiss, and L. C. Wall. 1958. High-Speed Cinephotomicrography of Human Spermatozoa. Med. Radiog. Photo. 34:44-49.

[63]   Rothschild, L. 1962.   Sperm Movement--Problems and Observations, In Spermatozoan Motility. D. W. Bishop, ed. AAAS, Wash-

ington, D. C. , p. 13-29.

[64]   Fray, C. S. 1971.   Observations on the Motion and Hydrodynamics of Mammalian Spermatozoa. M. S. Thesis, Department of
Mechanical Engineering, M. I. T.

[65]   Taylor, G. I. 1951. Analysis of the Swimming of Microscopic
Organism. Proc. Roy. Soc. A209:447-461.

[66]   Lamport, H. 1940.   Periodic Changes in Blood Estrogen.
Endocrinology. 27:273-680.

[67]   Thompson, H. E. , J. D. Horgan, and E. Delfs. 1969. A Simplified Mathematical Model and Simulations of the Hypophysisovarian
Endocrine Control System. Biophysical J. 9:278-291.

[68]   Schwartz, N. B. 1968.   New Concepts of Gonadotropin and
Steroid Feedback Control Mechanisms. Textbook of Gynecologic
Endocrinology. J. J. Gold, ed. New York: Hoeben.

[69]   Schwartz, N. B. 1969.   A Model for the Regulation of Ovulation
in the Rat. Rec. Prog. Hor. Res. 25:1-43.

[70]   Schwartz, N. B. and P. Waltz. 1970. Role of Ovulation in the
Regulation of the Estrous Cycle. Fed. Proc. 29:1907-1911.

[71]   Vande Wiele, R. L. , J. Bogumil, I. Dyenfurth, M. Ferin,
R. Jewelewicz, M. Warren, T. Rizkallah, and G. Mikhail. 1970.
Mechanisms Regulating the Menstrual Cycle in Women. Rec. Prog.
Hor. Res. 26:63-103.

[72]   Vande Wiele, R. L. 1970.   Private communication.

[73]   Speroff, L. and R. L. Vande Wiele. 1971. Regulation of the
Human Menstrual Cycle. Amer. J. Obs. and Gyn. 109:234-247.

[74]   Shack. W. J. , P. Y. Tam, and T. J. Lardner. A Mathematical
Model of the Human Menstrual Cycle. To appear in Biophysical J.

[75]   Stetz, C. 1971.   A Mathematical Model of the Human Menstrual
Cycle. B. S. Thesis, Department of Mechanical Engineering, M. I. T.

[76]   Pincus, G. 1965.   The Control of Fertility. New York:
Academic Press.

[77]   Rondell, P. 1970.   Biophysical Aspects of Ovulation. Biology
of Reprod. Suppl. 2:64-89.

[78]   Rodbard, D. 1968.   Mechanics of Ovulation. J. Clin. Endocrin.
Metab. 28:849-861.

[79]   Reproduction Research Information Service. (Bibliography of
Reproduction): No. 30, Mechanism of Follicle Rupture in Mammals.
1970.

[80]   Lardner, T. J., W. J. Shack and P. Y. Tam. 1970. A Note on
the Mechanics of Ovulation.   Unpublished paper.

[81]   Sawin, C. T. 1969.   The Hormones, Endocrine Physiology.
Boston: Little Brown Company.

[82]   Reproduction Research Information Serivce. (Bibliography of
Reproduction). 1969. No. 21, Sperm Transport in Uterus and Oviduct
of Mammals and Birds. 1970.   No. 26, Sperm Transport Through the
Cervix in Mammals.

[83]   Sobrero, A. J. 1963.   Sperm Migration in the Female Genital
Tract, In Mechanisms Concerned with Conception. C. G. Hartman,
ed. New York: MacMillan.

[84]   Hartmen, C. G. 1962.   Science and the Safe Period. Baltimore:
Williams and Wilkins.

[85]   Woodruff, J. D. and C. J. Pauerstein. 1969.   The Fallopian
Tube. Baltimore: Williams and Wilkins.

[86]   Rubenstein, B. R., H. Strauss, M. Lazarus, and H. Hankin.
1951. Sperm Survival in Women. Fertil. Steril. 2:15-19.

[87]   Moghissi, K. S. 1968.   Human and Bovine Sperm Migration.
Fertil. Steril. 19:118-122.

[88]   Moghissi, K. S. 1969. Sperm Migration in the Human Genital
Tract. J. of Reproductive Medicine. III:73-85.

[89]   Moyer, D. L., S. Rimdusit, and D. R. Mishell, Jr. 1970.
Sperm Distribution and Degradation in the Human Female Reproductive Tract. Obstet. Gynec. 35:831-839.

[90]   Van Demark, N. L. and R. L. Hays. 1954.   Rapid Sperm
Transport in the Cow. Fertil. Steril. 5:131-137.

[91]   Fromm, E. 1970.   Telemetering Systems for Long Term Observations of Uterine Contractility in the Unrestrained Animal, In
Biology of the Blastocyst. J. R. Blandau, ed. University of Chicago

Press.

[92]   Chow, W. W., A. G. Hardee and K. Millsaps. 1969. Two Simple Calculations on the Gross Movements of Human Spermatozoa on the Cervical Surface. Math. Biosciences.5:461-469.

[93]   Davajan, V., R. M. Nakamura, and K. Kharma. 1970. Review: Spermatozoan Transport in Cervical Mucus. Obstet, Gynec.Survey. 25:1-43.

[94]   Parker, G. H. 1931. The Passage of Sperms and of Eggs through the Oviducts in Terrestrial Vertebrates. Phil.Tran.Roy. Soc. London. B219:381-419.

[95]   Siegler, A. M. 1970. Observation of Uterine Contractility by Cinehysterography and Television Fluoroscopy. J. Reprod.Medicine. 5:115-122 (editorial page).

[96]   Cibils, L. A. 1967. Contractility of the Nonpregnant Uterus. Obstct. Gynec. 30:441-461.

[97]   Filler, W. W. and W. C. Hall. 1970. Dysmenorrhea and its Therapy: A Uterine Contractility Study. Am.J.Obs. and Gyn. 106: 104-109.

[98]   Lipner, H. 1971. Private Communication.